10 Essentials for High Performance Quality in the 21st Century

10 Essentials for High Performance Quality in the 21st Century

D.H. Stamatis

CRC Press
Taylor & Francis Group
Boca Raton   London   New York

CRC Press is an imprint of the
Taylor & Francis Group, an **informa** business
A PRODUCTIVITY PRESS BOOK

CRC Press
Taylor & Francis Group
6000 Broken Sound Parkway NW, Suite 300
Boca Raton, FL 33487-2742

© 2011 by Taylor & Francis Group, LLC
CRC Press is an imprint of Taylor & Francis Group, an Informa business

No claim to original U.S. Government works

Printed in the United States of America on acid-free paper
Version Date: 20111213

International Standard Book Number: 978-1-4398-7600-8 (Hardback)

Library of Congress Cataloging-in-Publication Data

Stamatis, D. H., 1947-
    10 essentials for high performance quality in the 21st century / D.H. Stamatis.
        p. cm.
    Includes bibliographical references and index.
    ISBN 978-1-4398-7600-8 (hardback : alk. paper)
    1. Leadership. 2. Time management. 3. Teams in the workplace. 4. Total quality management. I. Title. II. Title: Ten essentials for high performance quality in the 21st century.

HD57.7.S7153 2012
658.4'013--dc23   2011045393

Visit the Taylor & Francis Web site at
http://www.taylorandfrancis.com

and the CRC Press Web site at
http://www.crcpress.com

To Dr. Stephen and Kristina

Contents

List of Figures

List of Tables

Preface

The "future shock" once predicted by author Alvin Toffler is here. Those who prosper in our inconstant world accept change as a positive force and can persuade others to do the same. Those fearful of change or unable to rally others in new directions will be steamrollered with the volatility of modern organizations. You can get others to go along with useful change with a minimum of resistance. In the field of quality, we are no different. We continue to struggle to find a way to maximize product or service quality in relationship to cost and customer satisfaction.

But what is quality? The term *quality* has been with us for a long time. Ancient marvels have been constructed, as well as modern technology, all based on some of its principles. However, what is quality and how does it help the product and or service in the marketplace? Many definitions have been given for quality over the years; for example:

- Continual improvement (Deming 1982, 1986)
- Conformance to requirements (Juran, Gryna, and Bingham 1979)
- Quality is free (Crosby 1979)
- Loss function (Taguchi 1986–1987)
- COMFORT (Stamatis 1996)
- Six Sigma (Harry and Schroder 2000)
- And many others

Whereas the definition is somewhat easy to pinpoint, the application of quality in the marketplace seems to be nebulous if not very difficult to articulate. For example, a recent (November 4, 2010) Google search brought forth the following:

- Quality in relation to Malcolm Baldrige (684,000,000 hits)
- Quality in education (423,000,000 hits)
- Quality in religion (65,600,000 hits)
- Quality in health (513,000,000 hits)
- Quality in manufacturing (132,000,000 hits)
- Quality in nonmanufacturing (38,500,000 hits)
- Quality in service (901,000,000 hits)

From these statistics, we can see that the proliferation of quality has had a tremendous growth in awareness as well as implementing it to a variety of activities. However, the question remains: Why is it that organizations still have problems with their products and services? There are many possible answers to that question; however, we believe that the three fundamental ones are lack of understanding what quality is, appropriate quality planning, and applicable execution. As a society we still reward firefighters—that is, problem solvers—as opposed to planners—that is, individuals who avoid the problems to begin with. In other words, we still reward *appraisal activities*—after-the-fact occurrences—as opposed to focusing on eliminating these occurrences—*preventing activities*—from happening in the first place.

In this book we are going to attempt to give answers to these items by discussing and evaluating the core requirements of quality efficiency and improvement for a given organization in the twenty-first century. We are not going to spent time on identifying individual tools because there are many books that have been written about them. Specifically, we are going to address the specific issues (we call them essential) that any leader in any organization must follow to optimize improvement in their quality and customer satisfaction. *Improvement* by definition implies change—a change that will propel any organization into the world of excellence. This change, however, must be managed in order to be successful. A typical model for this may be the following steps:

- Appoint a champion for the change process who will build the right team to outline, facilitate, and implement the direction of the methodology for change. The champion will be someone who is highly respected, energetic for change, and has a track record of getting things done.
- Identify the values and beliefs that underpin the current organizational culture. Become aware of the impact of the prevalent management style, control structures, and methods of gauging real efficiency and effectiveness. Differentiate between *context* versus *content* and apply it appropriately. For example: Japanese focus on content. They will not stop until they have fixed a problem. Cost is a secondary issue. Americans, on the other hand, focus on context and therefore cost is a primary concern at the expense of fixing a problem. Content is how the product looks and feels to the customer, whereas context is engineering performance verifiable with data.

- Outline what the organization will look like at the end of the change program and build the case for change. This means getting people to recognize the need for change and getting them to own it.
- Define the scope of change in terms of markets and customers, products and services, business processes, people management and reward systems, structure and working relationships, and technologies.
- Assess the strengths and weaknesses of the management team to ensure that commitment is matched by competence.
- Draw up an outline plan covering the vision—what the organization is trying to achieve in simple, clear, brief and compelling terms; the scope—what the organization needs to change; the time frame—a schedule of activities; the people who will be mostly affected and who will play prominent roles (the change agents); resources—how much the change process will cost; the communication process—how the process will be communicated, reported and corrected; training requirements—in both hard (technical) skills and soft (interpersonal) skills; and structure—working and reporting relationships.
- Identify potential obstacles and restraining forces such as industrial relations practices and cultural propensity to change and learning.
- Communicate the what, why, how, who, where, and when to everyone concerned. Do not allow time for rumor or guesswork.

To be sure, change is inevitable and always present. Therefore, in order to make sure that quality improvement in any organization continues into the future of social and technological innovations and practices, we have isolated ten essential items—each in its own chapter—that will help the reader understand and implement the necessary improvement initiatives under the umbrella of quality. Each chapter is summarized as follows:

Chapter 1: The need for quality performance. In this chapter we consider the concept of quality performance and address some issues in specific sectors of our economy.

Chapter 2: Leadership. Here the focus is on leadership because it defines the path for improvement and it carves out the definition of success. The role of leadership is by far the most important ingredient in quality performance and customer satisfaction. In this chapter we focus on the characteristics of leadership and some of their roles and responsibilities.

Chapter 3: Execution. If leadership is the key component of performance, execution is a very close second. Without execution nothing happens. The best leaders are defined by their execution of their plans. This chapter focuses on project management, risk, and error analysis.

Chapter 4: Innovation. Innovation is the lifeblood of any organization's existence and the way to a prosperous future. Without constant, appropriate, and cost-effective innovation, the organization will fade away. In this chapter we will address design, manufacturability/assembly, and service.

Chapter 5: People and quality. Everyone knows that all organizations are made of individuals. Therefore, the ways in which we recruit, cultivate, and retain those individuals are very important. In addition, because the cost of recruitment is very high, it is imperative that we understand that our policies are not for the short term but rather for the long term. In other words, when we recruit we must realize that retention is important and is associated with consistency and future continuity of our policies. Specifically, we address the issues of empowerment, loyalty, multicultural management, business ethics, and training.

Chapter 6: Time management. Unfortunately, there are only 24 hours in a day. Yet, most of us have no clue how to prioritize events so that our scheduled activities are made according to plan. In this chapter we will address some of the basic issues of project/time management and its effect on the delivery of management's expectations or contractual agreements.

Chapter 7: Engagement. It has been said that management is the art of accomplishing things through others. To do that, the management team must be able to communicate what needs to be done, when, how, and where and in some cases why it has to be done. Therefore, in this chapter we are going to address the issues of communication, delegation, commitment, risk, managing change, and ownership.

Chapter 8: Team building. A team is the workhorse of activities in any quality-oriented environment. Without it very little is accomplished. In this chapter we will address some of the basic items that all teams must be aware of and the process of solving problems.

Chapter 9: Quality operating system (QOS). This chapter attempts to bring together a standardized way of looking at quality in any organization. As such, it provides a generic model to follow and two examples to show how this QOS model may be carried out in a work environment. The first example uses a baseball team and the second uses a specific automotive final assembly.

Chapter 10: Advanced quality planning. In this final chapter we provide an overview of planning and its effects on quality. We emphasize the notion of planning as opposed to appraisal methodologies to improvement so that improvement may be optimized throughout the organization and the supply base.

Epilogue: We close the book by recommending that quality is indeed a current issue in everything that we do, and unless we treat it with boldness, truth, and integrity we will not be successful in the years to come.

This work also includes six appendices, which are located in the companion CD-ROM:

Appendix A. Introduction of the International Standards (ISO 9000 and ISO 14000) and a very brief summary of the Malcolm Baldrige award.

Appendix B. Introduction of three major methodologies: (a) total quality management (TQM): specifically, we review the statistical process control (SPC) and some selected tools; (b) Six Sigma: specifically, we will introduce the Define, Measure, Analyze, Improve, Control (DMAIC) and the Define, Characterize, Optimize and Verify (DCOV) models; and (c) we will discuss the basic principles of Lean.

Appendix C. Root cause analysis. Problems occur in everything we do. Therefore, in this appendix we will discuss the importance of (a) root cause analysis and (b) corrective action.

Appendix D. A typical review for some analyses.

Appendix E. Opening doors of communication.

Appendix F. Time management tips.

REFERENCES

Crosby, P. B. (1979). *Quality is free*. New York: McGraw-Hill.

Deming, W. E. (1982). *Out of crisis*. Cambridge, MA: MIT CAES.

Deming, W. E. (1986). *Out of the crisis*. Cambridge, MA: Massachusetts Institute of Technology, Center for Advanced Engineering Study.

Harry, M. and R. Schroder. (2000). *Six Sigma: The breakthrough management strategy revolutionizing the world's top corporations.* New York: Currency.

Juran, J. M., F. M. Gryna, and R. S. Bingham. (1979). *Quality control handbook*, 3rd ed. New York: McGraw-Hill.

Stamatis, D. H. (1996). *Total quality service.* Delray Beach, FL: St. Lucie Press.

Taguchi, G. (1986–1987). *Introduction to quality engineering.* White Plains, NY: Asian Productivity Organization.

Acknowledgments

All books are a cumulative effort of many individuals and experiences. However, all books are also the result of certain individuals who have a tremendous influence on the final product. This book is no different. For me to identify all of the people who have helped and contributed in the final format of this work is an impossible task. On the other hand, there are some individuals who stand out, and without their help and support this book would not be a reality.

First and foremost I want to thank Drs. R. Munro and E. Rice for their moral support of this project and the patience that they showed when I was bouncing ideas about the content of the book. Their constructive criticism has proven quite worthwhile.

G. Mooridian is another person who, with his eloquent style of presenting an argument, made me think more specifically about the things I wanted to cover in this book. He was the force who helped me summarized my points as opposed to continuing to talk about them.

Dr. R. Kapur was invaluable in contributing thoughts and ideas about the future of quality in the twenty-first century. The many conversations we had proved very valuable for some of those ideas in this book.

Thanks are due to the Copyright Clearance Center on behalf of the *Training and Development Journal* for granting me permission to use portions of "Retention Tension: Keeping High-Potential Employees" by A. Nancheria and "Fix the Disconnect between Strategy and Execution" by M. Lippitt.

Thanks are due to the Copyright Clearance Center on behalf of the *Quality Digest* for granting me permission to use portions of "Fellowship: An Essential Element of Leadership" by P. Townsend and J. E. Gebhardt, available at http://www.qualitydigest.com/dec97/html/townsnd.html.

Thanks are due to the *Financial Times* and Mr. M. Christopher for granting me permission to use portions of the article "Coping with Complexity and Chaos," published on September 15, 2005, and available at http://www.ft.com/intl/cms/s/1/28d3583c-25eb-11da-a4a7-00000e2511c8.html#axzz1TK1U9Xgr. Furthermore, I want to thank them for allowing me to use some portions of the article by Mr. Ian Morley "When You Hear 'New Paradigm' Head for the Hills," published on June 12, 2008, and available at http://www.ft.com/intl/cms/s/0/e8e7e73c-3894-11dd-8aed-0000779fd2ac.html#axzz1SkSMES4p. Many attempts were made to find the original author, Mr. Morley; however, he has since left the *Financial Times* and there is no forwarding address for him.

Thanks also to Mr. John Tobin and the Design Management Institute (DMI) for giving me permission to use some material from the article "Radical Service Innovation" by Mark Jones and Fran Samalionis, first published in *Design Management Review* (Vol. 19, No. 1) and available at http://www.ideo.com/images/uploads/news/pdfs/08191JON20.pdf.

Thanks also to B. G. Wilson, D. H. Jonassen, and P. Colefor giving me permission to use some of the material from the article "Cognitive Approaches to Instructional Design," available at http://carbon.ucdenver.edu/~bwilson/training.html.

Thanks to Mr. Bill Wortman and the Indiana Quality Council for giving me permission to use some of their material on leadership (organizational structure; change management) and management skills (principles of management) from the *CQM Primer*, Fourth Edition.

Thanks are also due to the editors, who made a rough manuscript into a book that looks very professional and easy to read.

Finally, my most important supporter in this effort—as always—has been my chief editor, motivator, and life partner, my wife, Carla. She was always there to help me with the language, punctuation, and general editing. Her diligent effort has made this book more readable and easy to follow. Thanks, Carla Jeanne.

Introduction

If we always do what we always did we will always get what we always got!
This is a statement that many can identify with, yet we continue to be
resistant to change. The status quo is indeed very powerful and it is indeed
very difficult to change. But change we must, because it is "the" change
that brings hope for improvement, efficiency, ultimate customer satisfaction, and loyalty.

In the quality world, there have been many initiatives to reduce variation
and minimize cost, but every one of these initiatives fall by the wayside
after a while, and the problems continue to be fundamentally the same as
always. It is important to notice that in a survey of the quality literature
one will find the same old problems recurring even though we keep on
introducing new tools and methodologies for solving these problems. The
problems most often reported in the literature are lack of the following:

- Appropriate and applicable training
- Following procedures and instructions
- Effective communications at all levels
- Consistent and complete management systems
- Appropriate human engineering practices for the intended task
- Appropriate and applicable work flow for intended task
- Appropriate and applicable quality control/assurance practices

One hopefully will notice that these problems are not very difficult to
solve. However, they are critical in any successful organization for delivering success and meeting the goals of high performance. A close analysis
will help us understand why these items are not that difficult to fix. In fact,
one may even propose some self-explanatory fixes such as the following:

- Appropriate and applicable training
- Questioning attitude and intent, stop when unsure
- Effective communications at all levels
- Standardized procedure use and adherence
- Self-checking (responsibility at the source)
- Verification practices (accountability at the appropriate level)

- Pre-job brief (trial/pilot run)
- Peer-checking

Implementation of these will contribute to very good human performance because most of the items identified are indeed the majority of the contributors to our everyday problems. However, for true change to occur, the leaders of the organization must have the courage to act. That means the courage to challenge the way work is performed; to ask questions that people hope will not be asked; to point out ways in which management can improve the way the facility and or the organization is managed; to take a risk; to look for a different path for conducting business; and to find a way to make yourself an ally of those you think must change.

One may argue that the Sarbanes–Oxley Act of 2002 was introduced precisely to facilitate this challenge to ask questions without the fear of retaliation and the need for internal controls in the financial community. One of the concerns was to protect the "whistle blower." However, in reality, since the introduction of the bill we have seen some disappointing results. Out of 947 cases 70 percent have been dismissed, 15 percent settled, 13 percent withdrawn, and 2 percent have proceeded to administrative judges. Of those reaching the appeal stage, three have been reversed in favor of the company, two settled, and one remains open (Taub 2007).

To have an effective change, the objective of management is to help others see the same vision that is so obvious to them. However, management must invest time in identifying how the current problem is viewed and then help everyone arrive at the same conclusion that you now see as obvious. Only then, when working as a team, rather than as adversaries, does the chance of success become much higher.

To get everyone on the same wavelength is not as easy as it may sound. People see problems from different perspectives. An attorney may see the problem from a "blame" perspective. That is, who is to blame for this problem? The focus is *legal liability*. In real terms that means who pays and how much! On the other hand, a quality assurance engineer may see the problem from the perspective of "what" and "why" it happened. The issue of blame is secondary at best. The focus is to fix the problem in such a way that the information gained will help prevent this particular problem from happening again. We are not suggesting that the attorney's perspective is wrong. After all, if the attorney is successful, the end will be that his client would not be at fault and, therefore, the legal liability would be minimal. We are presenting two diametrically opposed examples to make the point of different perspectives.

The success of an organization must be viewed and evaluated not on the traditional model of social responsibility versus profitability but on a new model emphasizing social responsibility *and* profitability. This can be achieved by rejuvenating the concept of quality for high performance. We must be able to infuse multiple methodologies—as appropriate and applicable—to organizations so that waste (variation) can be reduced to a level that is compatible with the risk–benefit analysis for a given organization.

The new model is not an impossible task. In fact, many companies are very successful following criteria of quality performance in diverse areas, such as the following:

- Community
- Corporate governance
- Diversity
- Employee relations
- Environment
- Human rights
- Business ethics
- Product/service quality

This new model should focus on five basic functions, which are as follows:

1. Learning solutions: Formal, work based (coaching, collaboration, job aids, knowledge-based and challenging work tasks).
2. Nonlearning solutions: Culture change, strategy change, process improvement, data mining, and talent management.
3. Effective metrics: Revenue, profitability, customer satisfaction, retention, and productivity.
4. Learning operations: Internal control capability, outsourcing, standardization, technology infrastructure, e-learning, and scaling.
5. Efficiency gains: Time to deployment, content development costs, success, delivery costs, travel costs, time away from job, and time to competence.

These are functions that all manufacturing and nonmanufacturing sectors may benefit from; obviously, some are more successful than others.

Typical sectors (this is not an exhaustive list) that may benefit from this new approach of looking at quality are the following:

- *Agriculture/Mining/Construction (AMC).* Organizations such as agriculture producers; mining, oil exploration, and extraction companies; construction companies; fisheries; and forestry.
- *Finance/Insurance/Real Estate (FIRE).* Organizations such as banks and other credit institutions, insurance companies, securities brokers, and real estate companies.
- *Government.* Federal, state, and local government organizations or agencies (except public education).
- *Health Care.* Organizations such as hospitals, clinics, doctors' offices, and home care companies.
- *Manufacturing (Durables).* Manufacturers of durable goods such as wood products, stone and glass products, fabricated metal products, machines and electrical equipment, and transportation equipment.
- *Manufacturing (Nondurables).* Manufacturers of nondurable goods such as food and beverages, apparel, textiles, plastics and chemicals, paper and pulp, rubber, petroleum products, and leather. Also includes printers, publishers, and refineries.
- *Services.* Organizations such as hotels; business and personal services; automotive repair companies; professional services; educational institutions (public and private); legal, social, and other consumer services.
- *Technology.* Organizations such as computer, electronics, and communications equipment manufacturers; software designers; telecommunications services; information technology services and consulting firms; and manufacturers of medical equipment and other precision instruments. Specifically, one may identify a variety of learning technologies that may in fact help in a variety of ways improvement, in general. Some examples are the following:
  - *Learning Technologies.* This, of course, is a general term for electronic technologies that deliver information and facilitate the development of skills and knowledge.
  - *Cable TV.* This is the transmission of television signals via cable technology.
  - *Computer-Based Training CBT (Text Only).* This involves any learning event that is text based and computer delivered.

- *CD-ROM.* This is a format and system for recording, storing, and retrieving electronic information on a compact disc that is read using an optical drive.
- *Electronic Mail (e-mail).* This is the modern way of exchanging messages through computers.
- *Electronic Performance Support System (EPSS).* This is an integrated computer application that uses any combination of expert systems, hypertext, embedded animation, and hypermedia to help a user perform a task in real time quickly and with a minimum of support from other people.
- *Extranet.* This is a collaborative network that uses Internet technology to link organizations with their suppliers, customers, and other organizations that share common goals or information.
- *Groupware.* This is an integrated computer application that supports collaborative group efforts through the sharing of calendars for project management and scheduling, collective document preparation, e-mail handling, shared database access, electronic meetings, and other activities.
- *Interactive TV.* This is a one-way video combined with two-way audio or other electronic response system.
- *Internet.* This is a loose confederation of computer networks around the world connected through several primary networks.
- *Intranet.* This is a general term describing any network contained within an organization. It is used to refer primarily to networks that use Internet technology.
- *Local area network (LAN).* This is a network of computers sharing the resources of a single processor or server within a relatively small geographic area.
- *Multimedia.* This is a computer application that uses any combination of text, graphics, audio, animation, and full-motion video.
- *Satellite TV.* This is the transmission of television signals via satellite.
- *Simulator.* This is a device or system that replicates or imitates a real device or system.
- *Teleconferencing.* This is a way for instantaneous exchange of audio, video, or text between two or more individuals or groups at two or more locations.

- *Virtual Reality.* This is a computer application that provides an interactive, immersive, and three-dimensional learning experience through fully functional, realistic models.
- *Voicemail.* This is an automated electronic telephone answering system.
- *Wide Area Network (WAN).* This is a network of computers sharing the resources of one or more processors or servers over a relatively large geographic area.
- *World Wide Web.* All of the resources and users on the Internet using Hypertext Transport Protocol (HTTP), a set of rules for exchanging files.
- *Trade.* Organizations such as retail and food stores; home furnishings and equipment stores; general merchandise stores; food and beverage facilities; apparel and accessory stores; building materials, hardware, garden supply, and mobile home dealers; and wholesale trade or distributors.
- *Transportation and Public Utilities (TPU).* Organizations such as power, water, and gas utilities; trucking and warehousing companies; airlines and railroads; water transportation companies; and parcel services.

Many corporations fall in the above categories. However, some of the most successful companies that have followed the above generic functions for general improvement with a high rate of success are the following:

- Green Mountains (coffee roasters)
- Hewlett-Packard
- Advanced Micro Devices
- Motorola
- Agilent Technologies
- Timberland
- Salesforce.com
- Cisco Systems
- Dell
- Texas Instruments
- And others

High performance quality is the way to achieve high performance in any organization by understanding the functions that make quality achieve ever higher plateaus of profitability, customer satisfaction, loyalty, and corporate citizenship. We already know that specific tools and methodologies do contribute to improvement, but they are not enough, as history has proved. For example the following results were reported in *Quality Digest*:

> Six Sigma, 53.6 percent; process mapping, 35.3 percent; root cause, analysis 33.5 percent; Lean thinking, 26.3 percent; benchmarking, 25.0 percent; problem solving, 23.2 percent; ISO-based standards, 21.0 percent; process capability, 20.1 percent; statistical process control, 20.1 percent; performance metrics, 19.2 percent; control charts, 19.7 percent; process management, 18.8 percent; project management, 17.9 percent; customer-driven processes, 17.9 percent; design of experiments, 17.4 percent; failure mode and effect analysis, 17.4 percent; mistake proofing, 16.5 percent; poka yoke, 16.5 percent; process reengineering, 16.1 percent; change management, 14.7 percent; total quality management (TQM), 10.3 percent; variation measurement, 10.3 percent; Malcolm Baldrige criteria, 9.8 percent; workflow analysis, 9.8 percent; decision making, 8.9 percent; trend analysis, 8.0 percent; setup reduction, 6.7 percent; knowledge management, 5.8 percent; work breakdown structure, 3.1 percent. (Dusharme 2003, p. 3)

Despite our scientific achievements, we have allowed ourselves to be held hostage by individuals who propose specific methodologies and/or tools to gain advantage as though they are the utopian solutions to improvement and profitability. They continue to sell them as though they have found the "silver bullet," even though they know and everyone else knows that no such thing exists. We have convinced ourselves that specific approaches will answer the problems that face us only because we have chosen to chase false gods who will save us from an impending disaster only because we have failed to understand the fundamental ingredients of quality.

Advocates of quality typically insist that a specific quality program requires the complete commitment of top management. That sets up unrealistic demands because of the inevitable trade-off between various aspects of the business. If there should be a total commitment of top management it is to total performance management (TPM; not to be confused with total preventive maintenance); that is, to TPM rather than to any particular quality program/methodology or any other individual factor contributing to business results.

TPM requires professional attention (i.e., allocation of resources, appropriately and applicably) to multiple aspects of the business and to performance improvement. Quality improvement is frequently a major part of performance improvement. But so are such matters as focusing on profitable customer segments; enhancing product and service functionality; understanding what target segments actually want; product and service innovation; targeting promotion so it has impact; and productivity improvement. Rarely is a firm in a position to give each of these all the attention it would wish.

All in all, quality has made important contributions (and will in the future) to any organization that has made the commitment for improvement. That improvement may be institutionalized through active application of appropriate and applicable methodologies and tools but, more important, a commitment to reducing waste in the entire organization rather than just in some internal departments.

To be sure, improvement is also a function of how management's styles and models are influenced by an organization's size, industry type, and competition. Obviously, newly formed, small organizations typically have a flat (horizontal) or matrix type of structure. In the small organization of less than, say, 20 people, each person has several functional responsibilities. For example, the shop superintendent may also be responsible for purchasing, quality, and maintenance. Organizations with approximately 30 to 100 people divide the functional responsibilities so each person is wearing only one or two hats. An organization of this size may increase levels of management to perhaps three or four layers, from the hourly employee to the CEO or president, with an organization structure that is between a flat and a tall (vertical) organization. It is of interest in here to note that the Malcolm Baldrige criteria consider an organization with less than 500 employees as a small organization.

When organizations have in the range of 100 to 1,000 employees, more layers of management are required and the functional responsibilities are more finely divided. There may be separate quality control and quality assurance departments. Further, the number of employees within any one department is increased, simply because there are greater numbers of total employees. There is an increase in the types of organizational structures that may be used, including tall, functional, product, geographical, matrix, team based, or other forms.

Organizations with 1,000 to 10,000 employees, possibly with worldwide representation, usually fit the tall, functional, product, or geographical structure. There may also be a mix of structures with sales and service functions geographically structured, engineering and manufacturing with functional or product structures, and certain products being developed using matrix or team structures. The reason the matrix or team structures develop in a hierarchical organization is because management realizes that the organizational structure itself becomes an impediment to rapid change.

Organizations with more than 10,000 employees will have structures that fit tall, functional, or geographical descriptions or a combination of these. Of course, there are exceptions to all of these structural types, and a team-based organization of 8,000 may be found or an organization of 30 employees having six layers of management. The number ranges are approximate, based on the author's experiences.

Just as size affects the structure that organizations will typically follow, the type of industry also affects the structure. When the functional tasks to be performed are repetitive and require a large workforce, with each person doing the same activity, the organizational structure will be flat (horizontal). The span of control can be very wide, because all of the employees receive the same instructions, and they require little feedback from their immediate supervisors.

For complex products, where nearly every employee performs a different activity, a greater number of management layers are needed. This can be described in terms of project management using a work breakdown structure (WBS), where progressively finer detail is required, as the objectives are disseminated from the top to the bottom of the organization. The result is a tall organization with functional responsibilities.

Competition is yet another element that adds an interesting relationship to an organization's structure. If a newly formed organization is entering a mature market, with established, large competitors, then a market niche is available to the new organization if it can (1) be more responsive to customer needs, (2) respond faster to market changes, or (3) offer products at lower prices due to lower overhead costs. The new organization will adopt a matrix or team structure, with everyone clearly focused on the objectives.

The large, established organization, with a tall, functional, geographical, or product structure, will first see the new competitor as a nuisance and ignore any threats. As the competitor gains market share, the large organization will rethink its position and will form a matrix or team, within the hierarchy, to focus on the products that are being threatened. Depending upon the size of the large organization, it may take an excessive amount of time to recognize or respond to the new entrant. Typical examples are the entrance of Wal-Mart in the competitor world of K-Mart and the entrance of Dell and Apple computers in the competitor world of IBM.

The other way in which an organization responds to competition is by creating a geographical structure. This structure, as mentioned previously, may be a mixture of sales and service functions geographically structured and the balance of the organization as a hierarchy or other structure. The incentive to create a geographical structure, such as a national company, with local distributors or outlets is to enable flexibility in response to local customer needs. Snow shovels would be expected to have a low sales volume in Florida but high sales in Wisconsin.

As we have implied, in today's world, as well as in the future, organizations must keep pace with ever increasing changes. The complexity of the business requires additional functions in order to do business and be competitive. Some of the internal functional responsibilities of an organization include the following:

- *Human Resources*: The human resources (HR) department is responsible for an analysis of the needs and training of the workforce, employee turnover analysis, absenteeism analysis, and attitude surveys. In addition, the HR department may recruit, select, and hire people for the organization. It is important to note that this is the old personnel department. The change to the new name was viewed—by some—as necessary because humans are considered a resource and therefore they can be managed accordingly.
- *Engineering*: As a support service, production engineering is the problem-solving arm of the company. The engineering department should be proactive (always searching) in its problem-solving activities. The planning of new equipment or processes is a must for this department.
- *Sales and Marketing*: It is up to sales and marketing to develop effective plans to identify customers and markets for the company's current products and services and to identify needs and wants for new products. They should be in contact with engineering in order to pass along customer ideas or desires. The development of a marketing plan is necessary to help guide not only the marketing efforts but the production plan as well.
- *Finance*: The financial department in many organizations includes the accounting department, but these may also be separate departments. The accounting function includes the month end statements and the profit or loss for the company. A standard cost system should be in place so that data can be collected and based against it. Of course, other accounting methods may also be employed such as the

cash approach or activity-based accounting. The finance function is involved with foreign currency exchange rates, finding sources of funding, and investing excess cash. The evaluation of capital project requests for equipment, machinery, or buildings will usually be performed at the corporate level. The lease and purchase of equipment may also be evaluated there. The coordination of budget forecasts for the next year can be a part of the finance department as well.

- *Research and Development (R&D)*: Research and development activities are critical for the future of the company. The customer is satisfied for a certain time span with the existing product or service, but eventually the customer will want a new and improved product. Interaction with the marketing function and customer is needed to create new ideas and products. Working with the marketing and engineering departments can help make concurrent engineering a reality and reduce the cycle time of development.

- *Purchasing*: The securing of the proper raw materials, at the right time and the right price, is a basic requirement of the purchasing department. They must find ways to reduce the number of corporate suppliers, without increasing the risk of shutting down lines due to lack of product. The forming of alliances and partnerships among suppliers and customers is a current and future concern.

- *Information Technology (IT) and or Management Information Systems (MIS)*: The IT or MIS function is a key ingredient in the factory of the future. Many companies have already exploited information technology. The time savings, cost reductions, and information can make a company very competitive in the marketplace. Some of the benefits of IT or MIS that companies may take advantage of include electronic data interchange with customers and suppliers, electronic e-mail for communications, barcoding for all products, data collection for analysis, use of personal computers (PCs), online order status, and real-time inventory.

- *Production Planning and Scheduling*: Production planning and scheduling is a department that helps to coordinate the flow of materials throughout the plant. It tracks the levels of materials and inventory, schedules the product, tracks the product, and informs customers and suppliers of progress.

- *Quality*: The quality department is but one function in the corporation. Its emphasis may be on coordinating the total quality effort of a company and directing the quality assurance activities.

- *Manufacturing*: The manufacturing activity is associated with companies manufacturing a product or products. Manufacturing takes designs from engineering and schedules from planning and makes, assembles, and tests the company products. For a service organization, this function is replaced by the personnel performing the service function.

- *Servicing*: In this context, servicing is related to manufactured or sold product but can also be servicing of client accounts for non-manufacturing companies. Servicing is responsible for fixing problems with the product and assuring that the customers are satisfied after the sale.

- *Safety and Health*: The safety and health department aids the company in complying with local, state, federal, and industry regulations. The best known safety agencies include the Occupational Safety and Health Agency (OSHA), state safety agencies, the National Fire Protection Association (NFPA), etc. These agencies impact the establishment of a safety program, safety committee, and special safety task forces for the company. Pertinent paperwork should be handled by the safety department to satisfy various governmental requirements. The interpretation of safety or NFPA regulations would be a part of this department. Safety training in the areas of hazard communications, lock-outs, tag-outs, blood-borne pathogens, etc., would also be provided.

- *Legal and Regulatory*: A legal department (or attorneys on retainer) may be necessary to handle legal matters, especially in the very litigious society of today. The screening of letters of agreement, terms of leases, terms of lease/buy leases, property purchase agreements, sale of property agreements, rights-of-way, tax abatements, economic impact grants, etc., are examples. The legal department would also review press releases or other statements the company makes as official communications with the public.

- *Product Liability*: In the manufacture of certain products, the possible legal ramifications vary according to the product. A wood pallet grinder designed to grind pallets into one-inch particles also has the capability to do harm to a human being. The liability of the manufacturer may extend until the end of the life of the pallet grinder. The legal department would want to limit the danger and legal exposure of this product. The theories based upon breach of warranty have a statutory basis in the Uniform Commercial Code. Product safety requirements and labeling laws not only protect the consumer, but also should reduce the liability risk to the company.

- *Environmental*: An environmental department, separate from safety and health, is desirable. The proliferation of new regulations makes this a very volatile and difficult field. Regulations can be negated. What was permissible in the past can be deemed a violation today. The impact of The Clean Air Act of 1970 is presently a concern, as is effluent water. The Environmental Protection Agency (EPA) has jurisdiction over many of the emissions from a company. Companies seeking ISO 14001:2004 also need to look at use of energy. Sustainability concerns are gaining importance worldwide.

- *Technology*: A technology department is a luxury that only the largest companies can afford. This department is capable of scanning the magazines, journals, trade shows, conferences, patent applications, and libraries looking for new products and technology. Such an arrangement offers a competitive tool to the company. The smaller companies must depend on their existing people to pick up on new developments. If indeed there is a technology department in the organization, generally it is also responsible for the knowledge of the organization such as patents, processes, and so on.

- *Cross-Functional Collaboration and Systems Management Theories*: In traditional, functionally designed organizations, segments of vital activities are captive within and across many departments. For any client order, each department has the responsibility to process its part of the order as efficiently and as effectively as possible. Various department heads are responsible for the activities within their department, which allows for good management controls and procedures, but no one owns the overall process and the results. The client order fulfillment process involves the sales, finance, engineering, customer service, production scheduling, manufacturing, quality, warehousing, and shipping departments. Each department will be doing its best to optimize its own efficiency and output per department goals. Of course, each step through the process will increase the wait and queue times. If sales have promised quick delivery times for any item, regardless of volume, manufacturing might be intent on maximizing machine usage and will hold up small orders, to be completed with other similar orders. The result would be a product that gets into the customer's hands past the promised delivery date, with high production efficiencies and an unhappy customer.

Functional departments develop strong functional mindsets and will approach problems differently than other functional units. It is difficult when departments speak the different "languages," or have conflicting goals. The functional specialists will tend to focus on departmental matters and to the immediate superior's goals, not to the customer or to the industry. This is the creation of the *suboptimization principle* or, as it is commonly known, the *silo* or *stovepipe* mindset. An example of such a conflict occurs when manufacturing is trying to meet the end-of-the-month shipments but quality is holding the shipments because of nonconforming product.

Obviously, management needs to resolve departmental conflicts using one or more of the following methods (Thompson and Strickland 1995):

- Make support activities contribute to the success of critical process flows
- Contain the cost of support activities
- Minimize the time and energy units spend doing internal business
- Align departmental goals to support organizational goals
- Position the related activities to report to one person
- Use coordinating teams
- Use cross-functional task forces
- Use dual reporting relationships
- Reduce informal organizational networking
- Encourage voluntary cooperation
- Create group performance incentive measures
- Insist on teamwork and interdepartment cooperation

On the other hand, Galbraith (1995) described the cross-functional efforts as a lateral coordination effort. Departments or functions at the same level (lateral) should be grouped together to produce the required output. The units are all interdependent on each other if the firm is to succeed. For example, Boeing's 777 project had 250 teams performing the work. The teams were integrated quite closely and communicated frequently.

So the question then becomes "How can we improve any subsystem integration?" Schermerhorn (1993) provided some tips for such integration. They are as follows:

- Rules and procedures: everyone understands what to do
- Hierarchical referral: coordination problems go to a common superior
- Planning: objectives and targets are known by everyone

- Direct contact among managers: face-to-face contact among managers
- Liaison roles: cross-trading of personnel to work in each other's units
- Task forces: people from different units on task forces
- Teams: people from different units on task forces
- Matrix organization: create a matrix structure for specific projects

Companies have many resources available to them. Strategic planning, integrated into the business plans, will help guide them down the road. Unfortunately, the outside environment is something that many companies cannot predict but perhaps something that they can accommodate. The outside environment includes the following factors and forces:

- Competitors
- Customers
- Suppliers
- Technological advances
- Consumer rights
- Social forces
- Local, state, and federal units of government
- Regulatory units of government
- International governments
- The U.S. economy
- The world economy
- Environmental concerns

No doubt it would be very difficult for any one individual or department to be on top of every new regulation or issue at any one point in time. Centralizing functions related to the external environment and then disseminating the information internally facilitates decisions made at the department level being consistent with organizational level goals.

On this issue, Ivancevich and Matteson (1996) pointed out that the new enterprise architecture is upon us. The pressure to become more customer driven and to manage laterally is causing managers to rethink the configuration of enterprises. The firm must become more flexible, breaking down the internal boundaries between functions. It must become boundaryless. The nature of teams will be to bring people together to solve common problems.

The lack of boundaries will reduce the existing "us against them" thinking that is so common in organizations. The attitude will be to foster a common identity with the firm. There will be a focus on the process rather than function. This will also remove the silo effects of departments. The result of these changes will be that the overall goals of the organization are more likely achieved.

REFERENCES

Clean Air Act (1970). 42 U.S.C. §7401 et seq. United States Environmental Protection Agency. Washington D.C.

Dusharme, D. (February 2003). "Six Sigma Survey." *Quality Digest*. p. 3.

Galbraith, J. (1995). *Designing organizations: An executive briefing on strategy, structure, and process*. San Francisco, CA: Jossey-Bass.

Ivancevich, J. and M. Matteson. (1996). *Organizational behavior and management*, 4th ed. Chicago: Irwin.

Sarbanes–Oxley Act. (2002). The Sarbanes–Oxley Act of 2002 (Pub.L. 107-204, 116 Stat. 745, enacted July 30, 2002).

Schermerhorn, J. R., Jr. (1993). *Management for productivity*, 4th ed. New York: John Wiley & Sons.

Taub, S. (2007). "Cheap talk or deaf ears." *CFO* July, 14.

Thompson, A., Jr. and A. Strickland III. (1995). *Strategic management: Concepts and cases*, 8th ed. Chicago: Irwin.

Uniform Commercial Code (UCC). (1952 – 2003). UCC Articles - Article 2 - Article 8 - Article 9. Cornell University Law School. Ithaca, New York.

1

The Need for Quality Performance

In the introduction we emphasized the need to change and discussed some of the key issues in the quest for quality improvement. In this first chapter we discuss the concept of quality performance and address some issues in six diverse specific sectors of our economy.

Quality has been an issue for a long time in many industries as well as service organizations. We are not so much interested in defining specific issues or problems but in demonstrating the need for quality performance regardless of the organization. We have selected the following diverse sectors of the economy to show this need.

MANUFACTURING

Why is manufacturing leaving the United States? And what can be done to stop the trend? Many say that the end of manufacturing in the United States is the natural and inevitable result of a global economy. They say that manufacturing, which is heavily labor dependent, will seek the cheapest labor. But this is not the whole story. Most manufacturing is as capital dependent as it is labor dependent. And with more automation every day, labor costs are less of a factor than they once were. Instead, I propose that four other factors are just as important:

- *The Cost of Expensive Regulations.* The U.S. regulatory burden, especially unnecessarily expensive safety and environmental regulations, is almost nonexistent in Third World countries.
- *The Cost of Taxation Policies.* The policies of redistributing income at the expense of corporations are a concern because some of that income should be reinvested in future technologies. Large

international organizations always find a way to redirect profits to countries with the lowest tax rates; for example, Ireland with a 12.5 percent rate. At least for U.S. companies this means that the excess profits are not used in the United States to invest but remain outside the country. The result is that the United States suffers in innovation and technological endeavors as well as employment.

- *Too Little Investment in Improvement.* U.S. manufacturers, in an attempt to cut costs, have failed to invest in problem-solving/preventing technology like advanced root cause analysis. Thus, problems that could have been solved to cut costs recur while manufacturers implement ineffective, wasteful fixes.
- *Equipment Unreliability.* The cost of unreliable equipment at facilities is an unrecognized source of expense that magnifies labor costs. If manufacturers had more reliable equipment, productivity would improve (people would not waste time waiting around for frequent repairs).

The solution for two of these problems is not difficult or expensive in relationship to the total cost that they present. They require political intervention and applicable laws that will benefit the overall economy. The last two issues can be solved or at least be minimized with appropriate and applicable managerial attitude, engineering know-how, and quality methodologies for identifying opportunities for improvement.

In the automotive industry, we all have observed a slippery road for quality, performance, and financial stability during the last several years. The sale of an iconic brand such as Chrysler was an attention-grabbing event, especially when its German parent in effect paid $650 million to unload it after paying $36 billion to buy it 9 years earlier. This less-than-zero valuation illustrates the deep trouble that Chrysler is in as the result of high labor costs and mounting liabilities, particularly to cover rising health care costs. For Chrysler's new owners, the private equity firm Cerberus Capital Management, negotiating those costs down inevitably is job one. But another recent headline underscores what might be jobs two and three—innovation and investment—not just for Chrysler but also for General Motors and Ford (together known as the Big Three).

The Big Three have completely overlooked the auto market for the past several years; specifically, what the customer wants. They keep on producing vehicles that are expensive, inefficient, and a style that the general public does not want. In addition, their Japanese competitors, like Honda and Toyota, have already introduced hybrids and the first hydrogen fuel cell car. Honda's announcement is particularly telling because history is repeating itself. Japanese companies were first to market gas–electric hybrids, and Toyota has become the world leader in this growing field. This happened despite two U.S. taxpayer-funded research programs, the Partnership for a New Generation of Vehicles in the 1990s and President George W. Bush's Hydrogen Fuel Initiative, each of which pumped more than $1 billion into research.

This did not occur because Detroit's labor costs are high (on average about $1,200/car). It did not occur because health costs were high (about $1,800/car) or because of pensions. It occurred because the Japanese carmakers take a long-view approach and are willing to nurture cutting-edge products. In the United States, on the other hand, the focus continues to be on quarterly profits.

The Big Three have similar problems across the board; however, many analysts view GM as the closest competitor to the foreign transplants, in both current products and what it is planning. However, Ford, Chrysler, and GM will never truly catch up until they invest as much in research and development as their rivals, learn how to anticipate consumer demand, and nimbly execute a long-term plan. Among their problems are the following:

- *Mileage myopia.* In an era of $3- to $4-a-gallon gasoline, the Big Three are stuck with too many gas guzzlers. Even after two years of price swings, they have not aggressively made fuel economy a winning issue, as they did with air bags a generation ago. A version of the Ford Escape is the only American hybrid with any significant market share. In the 2008 model year all of the Big Three introduced several high-mileage vehicles; however, the improvement was not as breakthrough as it was expected and certainly what the customers were anticipating in both styling and mileage increases. The opportunity for example to increase diesel consumption was not taken seriously as it is in Europe, primarily because in the United States there are old negative perceptions for its use such as noise and smell. Furthermore the hybrids and electric cars even though they started to infiltrate the market there was no real demand for them primarily because of high pricing and lack of understanding of the technology. Some research is being conducted but not enough and not at the level necessary.

    Here we would be amiss if we do not address the issue of the introduction of the electric car—especially the one introduced by GM. Though mileage and CO_2 emissions will improve, the price of the car will be quite expensive even with government credits. In addition, the offset in pollution will be in the energy-generating plants, who mostly depend on coal and not on clean environmental generating capabilities. Yet another issue with electric cars is the notion of replacing an expensive battery and the questionable availability of total electricity needed due to inappropriate and insufficient grid technology.

- *Too much emphasis on internal surveys as opposed to listening to the customer and looking at their competition.* I have always been skeptical about the value of organizational-climate surveys—annual surveys that ask employees (a) how they feel about working for the company, (b) what is good or bad about the company, and (c) how they can improve their business. The premise is that results from these surveys will help management create a happier, healthier workforce and lead to better business performance.

  At best, surveys can only provide a rough idea about how people collectively feel about certain aspects of an organization's culture. What is needed is a deeper understanding of what survey results really mean. For example, one popular survey question asks employees whether the company communicates a clear strategic direction. The theory is that if employees know where they are going, they are motivated and can pull together in the same direction. This is a classic question, especially for the "vision" of the company. So, what does a collective "no" mean? Does it mean that communicating a clear, strategic direction will help people focus and enhance business performance? Not if individual performance measures are misaligned or if translating the strategy into implementation means different things to different functions.

  What about empowerment? What if survey scores indicate that people feel that management does not give them the freedom to do their jobs? Is the solution to be more hands-off? Not if employees do not have the experience or knowledge to make good decisions on their own.

  Many surveys evaluate areas that they should not. Some surveys ask whether employees are satisfied with their salaries. A better way to assess salaries is to benchmark them against the marketplace and work from there. It is important to evaluate the results of organization surveys to understand their true implications for business performance. Otherwise, improving survey scores may not have any effect on business results (Chao 2008).

- *Myopia of customer concerns and innovation.* Innovation is definitely the way to go. However, it must be focused on customer insight, which should be determined by (a) customer need and (b) a specific problem to solve. A successful approach for either one should be to

- *Develop a real solution and deliver real benefits.* The "real" should be determined by asking fundamental questions: Did the customer need this? Are we solving a problem?

- *Do not brainstorm too early.* Before you embark on any brainstorming, make sure that you have enough information. That is, do we have enough preliminary research to justify the innovation? Is the problem widespread and needs fixing? By doing this prework you may find that something else is needed.

- *Innovation must fit specific brand and/or corporate strategy.* Unless the innovation fits the brand and the overall strategy of the organization, it will fail. The appropriate fit is essential in developing effective and compelling innovations.

- *Be as specific as possible to the problem that needs solution.* Unless the identification is specific, ideation of a nonspecific area will likely lead to a solution that will not satisfy the customer.

- *Manage risk.* Companies that do not innovate well incur great risk. The development of a risk management process is imperative for the modern organization. It must be a complete process on its own, always keeping the customer central to the plan. A typical process may involve current information, gaps from existing to future wants by the customer, a list of the features and benefits that the customer wants, a rigorous innovation screening test plan for acceptance and buy-in from internal decision makers, appropriate budgets, further timing for future development, evaluation based on predefined milestone dates, and benchmarks for evaluating competitive products.

- *Diversity.* The best ideas are usually those that come from people not close to the business or product under consideration. Therefore, get as many people involved within and outside the organization to give you ideas.

- *Know your research team.* There are many methods to introduce innovation, including mind mapping, ethnographic observation, TRIZ, R&D, think tanks, blog mining, trend watching, and others. No matter whom you choose to be on your team in the innovation process, make sure that they know their process and, above all, that they have a track record of success. Do not partner with anyone who is clever in presenting innovation but is using you as the experiment.

- *Manage internal politics.* Perhaps the most important element of innovation is managing internal politics. Is management committed to innovation? Is there a budget for it? Is there a champion for it? For innovation to be effective and lasting, management must be in the center of it. If they are not part of it, innovation will never take hold.

An interesting example of innovation that has applied innovation and creativity differently was reported by Chantapalaboon (2008). A company tried to think of a better way of building brand loyalty. They wanted to go beyond offering a great product and service. They came up with the idea of creative solutions beyond the obvious ones. Traditionally, such a perception was crafted among customers through marketing communications. But this company wanted to encourage staff to think creatively. So they worked on matching customer perceptions of the product and service with the experience that sales and service staff provide. The new process is as follows:

A customer walks in the showroom because she has the perception that this place offers creative solutions. She tells the staff what she wants. The staff offers solutions to her problem as well as some creative options. The staff may use one of the creative techniques, such as the basic three questions (3-Q), by asking "what," "why," and "what else?" Or the staff may use the SCAMPER technique (S: substitute, C: combine, A: adjust, M: modify, P: put to other uses, E: eliminate, R: rearrange) to conceive more options for the customer. Of course, the staff may apply other creative tools as well if appropriate and applicable.

Then, the staff will use the pluses, potentials, concerns, and opportunities (PPCO) techniques to tweak those options to be more practical and realistic. The staff and the customer will discuss the pluses and potentials of those options and then openly identify concerns and explore opportunities to resolve those concerns.

The customer now has more practical options. And, the more choices there are, the better solution. From the customer's point of view, we need to ask the following: Is the customer impressed with the product and or service? Will he talk to his friends about it? Will he be back to use it again? From company's point of view, ask: What do you think about this application of creativity? Do you think the company would be able to differentiate itself from its competitors? Do you think it will surge ahead in its field? Would anyone try our product and services?

- *Stop-and-go planning.* Detroit churns through models while the Japanese companies keep improving old names, such as the Camry and Accord. The Ford Taurus could be the ultimate example. Once the best-selling car in America, Ford starved it of resources and then pulled the plug. Now its new CEO has dusted off the old name and slapped it on the struggling Ford 500 (Jones 2007).
- *Too many models, too little identity.* Detroit's brands stand for a dizzying array of products, from compact cars to massive SUVs of varying quality and appeal, some targeted heavily toward the rental market. Then these Detroit manufacturers wonder why they cannot establish brand loyalty for either makes or models. Detroit has, to be sure, made impressive strides in closing the quality gap with foreign companies—especially Ford Motor Company's Fusion model. And given recent statements by the United Automotive Workers (UAW) at all three companies, it is clear that both management and labor realize the seriousness of their situation (Jones 2007).
- *Investing in the future.* The top Japanese automakers spend more than their U.S. counterparts on research and development, as well capital expenditures, as a percentage of revenue (Shunk 2010). For example, according to the latest data available and reported by Merrill Lynch (2005) we see the disparity of investment shown in Table 1.1.

But that's not enough. The Big Three need more cars that excite people. And they need to be viewed as equal or better in quality. That—not just cutting labor costs, health care costs, and pension costs—will ensure their

TABLE 1.1

Research and Development Spending

Company	R&D (%)	Capital Expenditures (%)	Total Expenditures (%)
Nissan	5.0	6.0	11.0
Toyota	4.0	6.5	10.5
Honda	5.5	4.5	10.0
Chrysler	3.0	5.5	8.5
Ford	4.1	4.1	8.2
GM	3.3	4.7	8.0

survival. They must persuade the general public to buy American vehicles. It is an issue of perception (an unfair one for sure, but that is reality). No amount of government bailout money or loans from the government will make them viable in the future.

On the other hand, hybrids, electric, and fuel cell cars will not make them viable either because of the high costs. The return on the investment for the customer (even with government subsidies) is not commensurable with the benefits to the individual customer, not to mention the repair costs if these cars need repair.

NONMANUFACTURING

In the twenty-first century, the world is becoming a global marketplace for all companies. Competition is intense and companies worldwide need to compete with an international mind-set. Furthermore, the service element in most product and service organizations today is fast becoming a key differentiation that drives customer satisfaction as well as loyalty. That means that organizations must move beyond just delivering a product or service. The goal of the future organization is to build customer loyalty.

Quality in the nonmanufacturing (service organizations) sector as perceived by customers can be defined as the extent of discrepancy between the customers' expectations or desires and their perceptions. There are certain critical dimensions of service quality that a high-quality service organization must possess. They are as follows:

- *Reliability*: The ability to provide the promised service dependably and accurately.
- *Tangibles*: The appearance of facilities, equipment, and frontline employees.
- *Responsiveness*: The ability and desire to serve the customer promptly and efficiently.
- *Assurance*: The perceived competence and courtesy of frontline personnel, which results in customer confidence and trust.
- *Empathy*: The demonstration of a willingness to understand and meet each customer's unique needs.

Reliability is largely concerned with service outcomes and is the most important dimension in meeting customer expectations. On the other hand, tangibles, responsiveness, assurance, and empathy are largely concerned with service process and are most important (especially responsiveness, assurance, and empathy) in exceeding customer expectations. If an organization encompassed all of these dimensions would it be enough? The answer is maybe not. These dimensions may fail to deliver if the systems in the organization are not robust and dynamic. There are certain very critical issues that separate the good from the best and need to be understood.

Is providing good quality service the prerogative of large organizations selling a premium product or service? Can organizations that sell a reasonably priced service get away with shoddy service? Not anymore. It is not just those customers who frequent five-star hotels who have the right to good service on the basis that they are paying for it.

Even the leading customer-friendly organizations today fumble because they fail to explore on a continual basis the variation in requirements of the customer that creeps in due to various factors. The factors that have a bearing are primarily two:

1. *Multilocation factors*: The systems that work well in one location are not replicated across all locations to ensure the same level of service. Customers used to a particular level of service feel let down and might think twice before going back to even the main location, which created that level of expectation.
2. *Cultural issues*: These determine the way in which a customer will react to a particular kind of service. The organization also has to keep in mind the people (especially the front line) who are on the inside and the impact of their beliefs on their actions.

An example may help here in understanding these two factors. One of the leading hotels has a cycle time of 2 minutes for taking the payment made by a credit card and returning the card back to the customer with the bill. In one of the locations this took about 15 minutes. In response to the reason for the delay, the customer got a stiff apology with a vague answer about the delay in verification—even though it was 12:30 a.m., and he was tired from a long day's journey.

In another hotel famous for its service worldwide another customer was let down. He found a hair in the omelet he was eating. When he reacted the waiter apologized for the mistake and informed the manager. The manager was busy and did not bother to come and talk to the customer but just shrugged to the waiter and walked away.

What was the problem here? Was there no planning for errors and how to recover from it? Or was it a lack of sensitivity on the part of the front-line staff, which may be due to lack of empowerment? Or was it a lack of trust in the systems of the organization? We believe that this is a failure of the people and the service process. We believe that the systems have not been designed:

- To plan for errors or problems.
- To empower people.
- For a good recovery.
- To replicate where required.
- For providing a memorable experience.

We believe that any organization that has the goal to excel can make it right to the very top in the minds and hearts of their customers if their systems have the above-mentioned features. It is critical to understand the implications of each of these factors.

A favored story that W. E. Deming used to tell and is published in Gitlow and Gitlow (1987) is the following: John and Mary reviewed their daughter Penny's report card and were not pleased. "Penny, we know you can do better than this. Your average is 75," said John. Mary said, "Yes, Penny, by next term we expect you to raise your average to 85. You'll just have to try harder." "Okay Mom and Dad," said Penny. "I'll try."

The next day when Penny got home from school she tried to do her work, but she needed help with her math and her parents were not around. In addition, a neighbor was playing his stereo so loud she could not concentrate and her little brother kept bothering her. When her parents came home from work, she asked if they could help her but they said they were too tired. The next day Penny asked her teacher for extra help. The teacher said she would be glad to help her after school. But Penny had to go home after school to babysit her brother. She wondered, "How can I ever do better in school?"

Poor Penny! She is a victim in this situation. Her parents think they are helping by setting a goal for her, but it is an irrational plea. They have done nothing to change the system to help Penny achieve the goal. She would like to do better in school and get an 85 average, but what has changed to allow her to do that? Nothing! If Penny's parents are serious about her improvement in school, they should talk with Penny about the problems that are getting in the way of her improvement and should help her work on alleviating those problems. Penny should be involved in the process, because she can tell her parents what the barriers are. Several problems need to be addressed before Penny can improve. First, her parents need to be willing to help her or get her help that is available. Next, if help is only available after school then alternative babysitting arrangements have to be made. Finally, John or Mary should talk to the neighbor so that Penny can study in a quiet environment.

This, like the hotel example, is a system problem, and the parents (management) have to take responsibility to help solve it. John and Mary should accept this responsibility because Penny cannot do anything about these barriers. She would like to succeed in meeting the goal her parents have set, but she is stuck in an impossible situation that will inevitably lead to depression, frustration, stress, anger, and worsening performance.

Table 1.2 shows the ranking of some of the best nonmanufacturing organizations in the world market in 2007 based on a survey conducted by *Entrepreneur International Magazine*.

RELIGION

There is virtually no business in which quality is unimportant, and there is virtually no business in which quality dominates all other factors. Product function, cost, timing distribution, promotion, asset development, and asset management are also critical factors. An organization that does not manage all of these factors in addition to quality cannot expect success.

Though there have been many successes in quality management, there are also many failures of attempted implementation strategies for a given quality program. In starting and failing, these attempts may have damaged the organization. At the very least, these strategies diverted resources from more productive use. Proponents of a quality program characterize these as a failure of implementation and commitment by management rather than a possible unsuitability in the intended approach to quality.

Religious organizations are no different. Religious proponents tend to be uncompromising. Yet management inevitably requires compromise, or at least a balancing of competing demands, each with some important

TABLE 1.2

2007 Rankings of Some of the Best Nonmanufacturing Organizations

America's Top Global Franchises		Fastest Growing Franchises		U.S. Home-Based Franchises	
Ranking	Name of Organization	Ranking	Name of Organization	Ranking	Name of Organization
1	Subway	1	Subway	1	Jan-Pro Franchising Int'l.
2	Dunkin' Donuts	2	Jan-Pro Franchising Int'l.	2	Matco Tools
3	Domino's Pizza LLC	3	Dunkin' Donuts	3	Servpro Industries Inc.
4	McDonald's	4	Coverall Cleaning Concepts	4	Chem-Dry Carpet Drapery & Upholstery Cleaning
5	UPS Store	5	Jazzercise Inc.	5	Budget Blinds Inc.
6	Mail Boxes Etc.	6	Jackson Hewitt Tax Service	6	Bonus Building Care
7	Curves	7	RE/MAX Int'l. Inc.	7	SeviceMaster Clean
8	Sonic Drive-in Restaurants	8	CleanNet USA Inc.	8	Jazzercise Inc.
9	InterContinental Hotels Group	9	Bonus Building Care	9	CleanNet USA Inc.
10	Century 21 Real Estate LLC	10	Liberty Tax Service		
11	Jiffy Lube Int'l. Inc.	11	Cold Stone Creamery		
12	Papa John's Int'l Inc.				
13	Cartridge World				
14	Liberty Tax Service				

impact on business performance in a context of constrained resources. (Here we are not talking about dogmatic or spiritual items but rather the business of religion and its management.)

So, how can quality initiatives help the religious organization? By at least the following initiatives in the area of their management:

- Emphasis on measurement and its use to monitor quality.
- Highlighting analytical tools and their use.
- Recognition that quality is a joint responsibility for the hierarchy of the religious organization and the followers, not only the responsibility of a certain group at the expense of another. (It is important here to reemphasize that we are talking about the nonspiritual activities. All spiritual activities belong to and are controlled by the religious hierarchy.)
- Recognition that quality is primarily built into the organization's processes, not simply a consequence of individual actions.

An example in which a specific quality initiative may be implemented is in the environmental area. The organization may work to foster ecological stewardship and sustainability within religious congregations, judicatory bodies, and other religion-based organizations. (An excellent example of this type of involvement is the active participation of the Ecumenical Patriarch of the Eastern Orthodox Church.) The religious organization may collaborate with state, national, and international environmental organizations to sponsor conferences, programs, training seminars, advocacy campaigns, publications, and projects on a wide range of environmental and public health issues. In addition, they may provide opportunities for educational programs reflecting on the spiritual dimensions of environmental protection as well as information about environmental health, sustainability, and environmental justice.

Yet another example of where quality principles may be utilized is in the area of quality life (QOL) within the congregation. Subjective and objective indicators of QOL may be used to test relationships with religious involvement, participation, and belief. For example, the interfaith discussions of different religious traditions and denominations, including Buddhism, Christianity, Islam, Jainism, and Judaism, may benefit from principles of quality just like the discussions of ecological, sustainability, and health issues.

Quality efforts are based on several convictions about what is essential for supporting and sustaining strong and vital congregations. The first is that the quality of pastoral leadership is critical to the health of congregations. Effective leaders know how to deal with individual differences in a positive way. They know how to build *trust*. They know how to *listen*. When well-prepared, thoughtful, imaginative, able, and caring priests/pastors lead congregations, the result is that these communities of faith tend to thrive and grow.

Theological education is absolutely pivotal, and seminaries play a critical role in preparing priests/pastors/rabbis/imams for their leadership in congregations. Theological schools engage students in an exploration of the wisdom of the specific religious tradition (Christian, Islam, Jewish, and so on) and train priests/pastors, imams/ulemas, and rabbis in how to bring biblical and theological insights to bear on contemporary issues.

There is also the issue of ecology of the institutions, including congregations, regional and national judicatories, colleges and universities, seminaries, independent agencies, retreat and conference centers, publishers, and other supporting organizations, which must work collaboratively in addressing challenges and in maintaining strong and vibrant religious traditions and communities.

Furthermore, it is imperative that religious leaders understand the fundamentals of financial responsibility. Although they do have professionals dealing with financial matters, it is their responsibility to set the tone for fund-raising and expenditures as well as to define programs for appropriate funding. That knowledge should be based on quality principles.

Major research projects must support these efforts and provide a solid portrait of twenty-first-century American society and church life. This base of information should enable priests/pastors/rabbis/imams and all religious leaders to make informed decisions about their ministries and the broader public to understand more deeply the role of religion in American life.

In summary, quality initiatives in any religious organization may support efforts to

- Deepen and enrich the religious lives of all concerned (American Christians, Jews, Islam followers, etc.), primarily by helping strengthen their churches, synagogues mosques, etc.
- Support the recruitment and education of a new generation of talented priests/ministers and other religious leaders.

- Encourage theological reflection and religious practices that recover the wisdom of various faiths for our contemporary situation.
- Support scholars and educators who seek to help the American people better understand contemporary religion and the role it plays in our public and personal lives.
- Strengthen the contributions that religious ideas, practices, values, and institutions make to the common good of our society.

EDUCATION

A *pearl* is something precious or highly valued, because a pearl represents purity, wisdom, and spiritual transformation. In the modern world of global competition, education is the pearl of global competitiveness. Many organizations spend millions of dollars to educate and train their employees to bring them in line with the new technological innovations.

Education has taken a new form in the sense that now we talk about "lifelong learning" for the commercial world and a new approach to formal education in offering and teaching methodologies from K–12 as well as university classes.

The problem with education in our modern world seems to be faced with conceptual and normative underpinnings of various views about the definition of education, educational quality, whether or not we should have it as a centralized governmental function, and what the role of education is in our society at large.

These are worth the effort of research and debate. However, unless we define the goal of education, we will be debating education for a very long time. Currently there are two trains of thought: (1) transcendent justifications of educational aims and (2) sociocultural reproduction justifications. In both cases we are talking about teleological results. *Teleology* comes to us from the Greek word *telos* which means "end" or "purpose." Teleology, then, is the philosophical study of design and purpose.

In education we must not only have the definition and the purpose up front but we must define the strategy to get us there. Case in point: The ancient Greeks had the educational goal of their society as "produce a good citizen." Everything they did was toward that end. They focused on the *arete*, which we now define as "excellence" or "virtue." They created heroes to support that endeavor via human characters such as Hercules, Achilles, and so many others. On the other hand, Western education philosophers have been concerned with the question of educational quality. The focus has been drifted from excellence to "what it means to educate others," or "to be educated well." In essence, educational quality has been defined not as a holistic endeavor but quite fragmented in the name of specialization. So now we have come to a point that our educational process has many goals instead of one goal. In other words, we have lost as a society our teleological existence and we aim aimlessly at education without even knowing what education is supposed to be.

A classic and disheartening situation is when we say that our goal is to be educated, implying that without education we would be worse off. But are we really better off? Are individuals who graduate from high school and college who cannot speak or write properly "educated"? When we force professionals after pursuing graduate work to take certification exams—because we allegedly want a standardized knowledge base—are they properly educated? What does it mean to be properly educated? This is the crux of the situation, and that is why there are so many ways to present education with no common substantial agreements as to what constitutes excellence.

Education must be changed. The change must be cultural, depending on what society needs, and scientifically based. In both cases the goal and strategy must be in the forefront of any change.

The brain has evolved in three stages: the *reptilian* brain (action oriented), the *limbic* brain (feeling oriented), and the *neocortex* brain (thought oriented). Scientists refer to this total brain development as *triune*, acknowledging the uniqueness of the three phases along with their overlapping, but separate, functions. The limbic brain first distinguished itself from the reptilian (primitive) brain with the advent of mammals. The limbic (sensory/emotional) brain takes in information from the senses and processes visually. It is where dreams occur. The limbic brain was in operation for millennia before the neocortex (learning) brain evolved. Language, because it is a system of abstract symbols, only became possible after the development of the learning brain. We are therefore hardwired to think in images, not words.

We do not—indeed cannot—use each step of mental processing every time we encounter a stimulus. It would simply be overload for the three-pound universe residing in our skulls. Looking at mental processing in a linear manner that begins with input and ends with rational decisions is misguided. It is the ability of our limbic brain to use visual cues as instantaneous and accurate guides that allows us to survive each day. For example, look at traffic signage worldwide. Fairly similar, these signs are able to impart valuable information to people without the use of language.

Furthermore, using verbal input as the basis for qualitative research runs into issues of context. Human beings, as social animals, must rationalize their reactions and compare against norms to decide whether or not what they want to say is acceptable. Facial coding bypasses this overanalyzed verbal input by getting true reactions that occur as the brain processes sensory cues. Eye-tracking tells us what these sensory cues are.

The traditional scientific view held that emotion came after rational thought processing. However, research into fear circuitry by LeDoux (1996) of the Center for Neuroscience at NYU suggests the opposite. He proposed that there are, in the emotional decision-making process, two different routes that we unconsciously choose in response to external events. Incoming sensory information first gets filtered through the thalamus, the screening device for the psyche, which evaluates the input for interest and relevancy. It then goes to the hippocampus, where anything of any emotional significance or reminiscent of familiar associations gets in; information deemed worthless never gets routed to either the conscious, learning brain or the sensory/emotion-based limbic brain.

Sensory input is routed to one of two paths: (1) The limbic path goes from the thalamus straight to the amygdala. It is more immediate, responding to sensory input prior to any conscious thought. If the sensory input has enough emotional kick to it, the limbic brain and the body work together, on instinct. They ensure that chemicals are secreted to heighten our alertness while the muscles are prepared for action. (2) In contrast, on the learning path the rational brain predominates. The sensory input filtered by the hippocampus and handled by the thalamus is then passed to the learning brain. There, it is more rationally analyzed and will be passed on to the amygdala if the input has any emotional importance.

Both paths lead to the amygdala, the brain's emotional thermometer, mobilizer, and short-term memory storehouse. The amygdala instructs the body to marshal its emotional resources and prepare for response. Sensory input channeled to the low road due to emotional urgency may activate a response even before the high-road learning brain has had a chance to perform its analysis (Hill 2006). On the other hand, even information processed by the frontal cortex and learning brain must go through the amygdala for action to be taken. No matter what path is chosen, the outcome is emotion-driven action.

This is probably the most important issue that will affect how future research is conducted. Current knowledge about the decision process has demonstrated that emotions significantly drive outcomes. Conversely, current research methods ask subjects to reverse the true nature of the thought process by asking them to rationalize (think) what they feel when in reality humans feel and then think. Emotion is the driving force behind decisions no matter which path the brain uses. Eye-tracking defines whether or not the correct message was delivered in the first place. Facial coding quantifies these emotional drivers and ascertains whether the intended message evokes the desired response. Together they offer the ability to create more effective emotional connections that leverage either path of the decision process by ensuring emotional importance and relevance.

In his book *How Customers Think*, Gerald Zaltman (2003) reported that cognitive scientists estimate that at least 95 percent of our thought processes are not fully conscious. Conscious thought is merely the tip of the iceberg. We are therefore much less in control of our decision making than we believe. For marketers, this means that appealing to consumers on a conscious, rational basis and asking them to evaluate features, attributes, and benefits is largely incorrect. To reach consumers and turn them into loyal buyers, you must appeal to their emotions. For researchers, this means that self-report scores such as those used in Internet surveys, clipboard questionnaires, and the like are not reliable. For quality professionals, it means that the product and/or service produced by the organization must meet the customer's requirements and in fact sometimes surpass them.

The issue at hand in current research methodology is asking subjects to consciously verbalize their internal thoughts and feelings. But if the subconscious controls most thoughts, it is technically impossible for most consumers to provide an accurate assessments of their thoughts, feelings, and reactions; they simply do not have access to the underlying cause and motivation behind their decisions. The main implication for research is that consumers simply cannot think what they feel; they can only feel what they feel. By avoiding complete reliance on verbal responses that are not necessarily relevant to internal feelings or thoughts, facial coding can deliver accurate measurement of true emotional response. It provides quantitative measures of a qualitative subject by simultaneously uncovering how a consumer feels and measuring the frequency of emotion present. Eye-tracking is able to mirror the subconscious journey that is taken visually. When the two methods are combined, they provide emotional insights that sync up with actual visualization patterns. In other words, no longer do you have to rely on subjects claiming that a certain thing caused their reaction; it is possible to know specifically what the catalyst was.

Two-thirds of communication is nonverbal. Most communication experts agree on this statistic, and the percentage jumps to 90 percent when the topic has some emotional weight to it. Humans developed the capacity for verbal communication relatively late in the game. The capacity for word-based communication came with the development of the neocortex—the most recently evolved part of the brain—which is capable of processing abstract thought. Until then we relied on sensory signals to assess and gestures to communicate. According to *The Silent Language* (Hall 1959), there are ten primary modes that humans use to communicate, and only part of one of those modes involves actual verbal language. Verbal communication is routinely influenced by nonverbal signals that we deem to be more credible.

By not taking nonverbal response into account, most current market research methods disregard almost two-thirds of human communicative potential. By focusing mainly on verbal and—as neuroscientific advancements have shown—not particularly reliable means of communication, research that is supposed to be qualitative in nature does not provide what it promises. Facial coding leverages this often unaccounted for nonverbal communication to create qualitative data that needs no support from subjects but instead relies on their internal desires and feelings.[1]

What does this purely scientific discussion have to do with high performance quality? Plenty! This scientific knowledge has a profound impact on how we proceed with lifelong learning. This means that our effort to optimize learning for workers is just as important as training the executives. The Novations Group did a study in early 2007 that showed that most executives are ambivalent about training because they have other things that they have to do. However, as the group pointed out, the problem is that there are many first-line managers who might be brilliant technicians, but when they get promoted to a senior role and suddenly they are out on their own, in a world where resources are squeezed, pressures intense, and the politics toxic, they find out the hard way that indeed they lack appropriate training.

It is imperative for organizations to be able to diffuse both organizational and new knowledge into their training and education development. Here we must also emphasize that the theory of the 5S's is working precisely because it is visual and it also communicates on the subconscious level.

As for formal education (K–12 and college), change is in the air. For the past several years education in the United States has been lagging behind several countries. In fact, U.S. students are continuing to lag behind their peers in other countries in science and math, as the test results published on December 3, 2010, showed. The test, the Program for International Student Assessment, was given to 15-year-olds in 30 industrialized countries in 2006. It focused on science but also included a math portion.

The 30 countries, including the United States, make up the Organization for Economic Cooperation and Development (OECD), which runs the international test. The average scores for U.S. students were lower than the average scores for the group as a whole. U.S. students also had an average science score that was lower than the average score in 16 other OECD countries. In math, U.S. students did even worse—posting an average score that was lower than the average in 23 of the other leading industrialized countries.

The test also was administered to students in about two dozen countries or jurisdictions that are not part of the industrialized group. When compared with the broader group, U.S. students fell in the middle of the pack in science and did somewhat worse in math. There was no change in U.S. math scores since 2003, the last time the test was given. The science scores are not comparable for the years 2003 to 2006, because the tests were not the same. U.S. girls and boys did about the same on the science and math portions of the test. Finland's 15-year-olds did the best on the science test, followed by students in Hong Kong and Canada. Students in Finland, Taiwan, South Korea, and Hong Kong were the top performers in math.

Obviously, something is wrong here and we must do something about it. There is dire need for innovation. This innovation has to be addressed using a holistic approach from at least six perspectives: (1) the need to revamp grades K–12, as far as both objectives and content are concerned; (2) a way of teaching that takes seriously the "why" question students often pose and revises and rethinks its own aims in the process of teaching and learning (it is not acceptable anymore to say: "Because it is on the test," "Because we have always done it this way," or, as a last resort, "Because I say so"); (3) introducing critical and liberating activities of the classroom itself that support the teaching points; (4) revamping the college undergraduate curriculum and its teaching methodologies (place accountability and responsibility on both faculty and students alike; introduce technology into the classroom and, in some cases, introduce education without the classroom [distance learning and online learning]); (5) reintroducing vocational education and training in the scheme of things; and (6) reevaluating the notion of lifelong learning and education for all.

It is imperative for all of us to understand that education and training are the cement that binds societies together in the face of economic and demographic change. It is too important to be left only to the politicians. It is a travesty for our society at large to send students to college and make them take remedial classes, especially English and math—subjects and content that should have been covered in high school. We waste time and money for things that should have been learned. However, there is no one to complain to and no one to take responsibility for that failure. Rather, we are cheerful and proud that our kids are in college.

It is time for the education community as well as parents and society at large to demand high-quality education with the least amount of time, effort, and cost. These are the primary ingredients that can make the education system an excellent one by using the principles and methodologies of established quality initiatives such as character education, total quality management (TQM), Lean, Six Sigma, and so many others.

TQM is the very basic approach for quality management to be implemented in our educational system (both formal and informal). At its core, TQM is customer driven. In other words, it is a system of principles that refer to a management process and set of disciplines that are coordinated to ensure that the organization consistently meets and exceeds customer requirements. In the case of education this is quite difficult, because the customer is defined in at least four ways: (1) parents, (2) students, (3) society, and, to some extent, (4) the government. The outcome of education, then, must be to satisfy or exceed the expectations of all of its customers.

Ironic as it may seem, the ancient Greeks defined education around three major subjects: (1) grammar, (2) rhetoric, and (3) logic. They regarded these as practical, useful, and indispensable skills that were seen as the foundation of preparing the youth for citizenship and participation in a free society. That preparation was sufficient for them to invent "science" as we know it today. To paraphrase Rhodes (2006), how doubly ironic, then, that in our science-driven age we have so little place for the wisdom of Greece. It is not that we reject useful knowledge. We worship it, but we have redefined it to exclude those very elements that the Greeks judged so significant. It is time that we reconsider our approach for the benefit—and perhaps even survival—of our modern world.

But Greeks were not the only innovators in reforming education. If we review the Middle Ages we will find that arithmetic, astronomy, geometry, and music were added to that core knowledge base, thus providing the essential educational preparation for the new age of discovery that marked the Renaissance through specialization.

It is our generation that has seen a wave of educational transformations into practical education. That means that our leaders, who may be engineers and architects, physicians and social workers, lawmakers and urban planners, business executives and economic policy makers, have graduated untouched by the hard-won collective historical experience, social perspectives, moral considerations, and humane reflections of our fellow human beings through the ages. Unencumbered by such reflections, they are likely to confront each new emerging issue as something novel: a challenge encountered by society for the first time (Rhodes 2006).

It is time for our formal educational process to perhaps establish in education the concept of *sustainability*, which could provide a new foundation for all education. Sustainability is a concept used in the environmental movement that means to preserve with minimum disturbance Earth's bounty—its resources, inhabitants, and environments—for the benefit of both present and future generations. The old Native American proverb captures perfectly the spirit of this sustainability! We do not inherit the earth from our ancestors; we borrow it from our children.

In terms of the educational process, how we can use the concept is a fair question. Certainly, some significant exposure to the appropriate sciences (e.g., mathematics, chemistry, physics, biology, geology, natural resources, ecology, climatology); understanding of social interaction (e.g., sociology, economics, history); extensive familiarity with the great issues and themes of human inquiry, self-reflection, and moral consideration that have guided human conduct and are reflected in human creativity (e.g., arts and the humanities); and, of course, technological issues (e.g., computers and information technology) are required. In other words, sustainability in education is building knowledge that is relevant and useful for the years to come. That building process is to learn what has worked and what has not and incorporate it into a dynamic system for success.

Though this may be interpreted by some as "same old stuff," in fact, it is different in the definition of a new focus, added coherence, and stark immediacy that it would provide. Beyond the complexities of sustainability the larger concern is "for what purpose?" It is our opinion that for sustainability to be understood in the context of education we must frame it in terms of value, meaning, and purpose as defined by our society at large.

Here, of course, we come back to the Greeks, who defined the purpose of education as creating good citizens.

Lean, Six Sigma, and many other improvement approaches are specific methodologies for improvement within the educational process. However, the introduction of character education as an infrastructure to the educational process all across the K–12 and college curriculum is essential and fundamental for good citizenship.

The character education label applies to a wide range of programs and strategies in schools and is designed to help young people become morally responsible, engaged citizens. The Character Education Partnership (CEP) promotes an intentional, proactive approach that affects all aspects of school life. Working with experts in the field, CEP has identified 11 broad principles for comprehensive and effective character education:[2]

1. Promote core ethical values as the basis of good character.
2. Define character comprehensively to include thinking, feeling, and behavior.
3. Use a comprehensive, intentional, proactive, and effective approach.
4. Create a caring school community.
5. Provide students with opportunities to engage in moral action.
6. Provide a meaningful and challenging curriculum that helps all students to succeed.
7. Foster students' intrinsic motivation to learn and to be good people.
8. Engage school staff as professionals in a learning and moral community.
9. Foster shared moral leadership and long-term support for character education.
10. Engage families and community members as partners in character education.
11. Evaluate the character of the school, its staff, and its students to inform the character education effort.

So why aren't all schools doing it? In the early history of public education, developing good character was seen as an essential part of preparing people for citizenship in a democratic society. But in the latter part of the twentieth century, many public schools moved away from the traditional emphasis on character and citizenship as American society grew more complex and diverse.

Today, character education is making a comeback. Thirty-one states mandate or encourage character education by statute. Though pronouncements by legislatures do not necessarily translate into quality character education programs, it is a start.

Much is at stake in getting this right. At this critical moment in America's history, we need far more than higher math and reading scores. We need citizens who have the strength of character to uphold democratic freedom and world competition in the face of unprecedented challenges at home and abroad. "Only a virtuous people are capable of freedom," is the familiar aphorism from Benjamin Franklin (Ellis 2008). Less well known, but worth recalling, is the warning in the sentence that follows: "As nations become corrupt and vicious, they have more need of masters."

HEALTH

Global trade and investment has surged in the past 50 years. Almost all countries and all sectors today participate in a global division of labor. Yet one sector remains conspicuously unglobalized: health care. Many governments around the world have ring fenced health care provision[3] and rejected virtually all attempts to open this sector for cross-border exchange. Resistance has been so fierce that the European Commission, for example, now hesitates to table a new directive that would simply codify a ruling from the European Court of Justice: to ensure free movement for patients and the right to be reimbursed for treatments abroad if the national health system cannot provide the treatment within a reasonable period of time. This proposal has been due since December 2007 and there was a new deadline by the end of 2008. As of this writing the directive seems far away; new versions constantly appear with ever more diluted free trade credentials.

Why is free trade in health care so resisted? Outdated notions of how to organize health care systems lie behind the core ideological opposition. False concerns for health care in developing countries have also become expedient handmaidens for health care protectionism in the developed part of the world. The truth, however, is that many developing countries want to liberalize trade in health care. In the World Trade Organization (WTO), it is developing countries that have made the strongest commitments. Countries as diverse as India, Cuba, China, Thailand, Jordan,

South Africa, and the Philippines have all developed export strategies to supply health services to foreign markets. They are knocking on the doors of Europe and the United States to sell their health care services. Similarly, patients from the developed world increasingly buy treatments in developing countries. In 2006, Singapore treated half a million patients from abroad. India claimed 600,000 foreign patients and Thailand as many as 1.2 million health care "tourists."

According to a study by McKinsey, the consultancy firm, medical tourism by 2012 could bring an annual $2.2 billion to India alone. This figure may not sound impressive considering the $3,000 billion value of global trade in commercial services. However, Indian revenues from export of software, services, and business process outsourcing—the much-talked-about success story of India—amounted to $20 billion as recently as 2004 (but grew to $31 billion in 2007) and export of health care is still only in its infancy. Reforms opening up trade would lead to much greater benefits. Western patients, and governments, could save huge resources by receiving treatments from developing countries. Modern technology, such as telemedicine—procedures carried out under the instruction of distant specialists—allows production networks to be created among hospitals in developed and developing regions (Indian Express).

Today health care represents only 0.2 percent of all foreign direct investment in services, but the price differences tell us something about the potential. Open heart surgery in India, for example, costs only an eighth of the price in the United States (travel, accommodation, and medicines included). A hip can be replaced in Thailand for only a sixth of the cost in the UK. These price differences should appeal to cash-strapped health care systems in the United States and Europe. In spite of high tax revenues in recent years, European governments have to ration health care to slow down rising expenditure. Hospitals lack resources to invest in new technology. Many new drugs are not covered by health insurance. But these policies to contain cost do not really work; mostly they target marginal spending but neglect the overall inefficient use of resources. Economy drives that prohibit the purchase of gene scanners or prevent reimbursement of the cost of a new cancer drug with limited side effects have a very small impact when health care productivity overall is falling. Furthermore, they cannot hold back expenditure as demand for health care is increasing (Erixon 2008).

Free trade in health care is no panacea, but it would certainly help all governments use resources more efficiently and improve accessibility and affordability of health care. Substantial liberalization is unlikely to emerge from WTO trade negotiations, at least in the foreseeable future. Governments should therefore opt for autonomous reforms, tailored to their needs, and stronger regional trade and investment cooperation.

Since about 1990 much has been talked about health care. Many initiatives in cost cutting and quality, including TQM, Lean, and Six Sigma, have been introduced with positive results that have actually improved the cost, as well as service, of health care, but much more is needed. For example: medical errors and medication errors in the United States continue to be a problem. This continues to be a problem in the face of procedures for analyzing mistakes made while using medical monitors and medical devices. Perhaps a solution would be a system of better human engineering of medical devices including better labels and better designs of displays and controls—at a very minimum.

The Institute of Medicine (IOM) has released another eye-opening report. This report calculated that each year there are 1.5 million medication errors that cause injuries in the United States. For example, a 1999 report by IOM estimated that there are as many as 98,000 deaths each year caused by all types of medical errors (Newmart 1999). These two reports suggest that the medical industry needs to adopt best practices to help solve the problems that cause all types of medical errors.

Since 1999, many discussions and conferences have been held throughout the country and most errors occur in the following areas:

- Serious discussion of health care's best and worst practices in need of detailed analysis
- Medical error case studies
- Stress and human error
- Stopping hospital infections
- Human performance lessons from major accidents
- Lessons from crime scene evidence preservation for accident investigation

One can see that unlike many wonks[4] who foolishly believe that health care is not a market, we see competition "of a sort" at work; namely, zero-sum competition that adds little value, fosters inefficiency and poor quality, and often harms patients. Why? Because the current competitive environment is dysfunctional; serves to "shift costs, accumulate bargaining power, and restrict services"; and is ultimately misplaced, focusing on the business dynamics of providers and health plans rather than on the diagnosis, treatment, and prevention of illness.

Focusing on how to move American health care to positive-sum competition based on economic and clinical value for patients, redefining health care provides a series of specific recommendations for the key players; notably, providers, health plans, employers, and Medicare/Medicaid policy makers.

Specifically, in order to change the traditional model we must begin to see health care as a business in need of innovation. For example, a way that:

- Provides an insightful, detailed overview of the most influential players; namely, medical professionals (medical doctors, administrators, and medical personnel) and the pharmaceutical, biotechnology, genomics/proteomics, medical device, and information technology sectors.
- Describes and assesses the market structures, business models, and corporate strategies of each of these six sectors.
- Shows how the six sectors are converging, drawing increasingly on the trends, tools, and solutions of each other.
- Focuses on quality and process improvements.
- Focuses on reduction of medical and administrative expenses, increasing collections of bad debt and overdue insurance payments.
- Introduces electronic medical records and electronic prescribing systems.
- Creates affiliations or joint ventures with physicians.
- Initiates collaboration with rival hospitals on high-priced technology and medical services.
- Refinances high-interest, variable-rate debt.

Medical facilities are adopting practices from other industries to prevent medical errors. One of the techniques that has been adopted by some is called SBAR. SBAR is a communication technique used by the military to help verbally turn over information from one watch officer to another. Hospitals are adopting SBAR for patient handoffs (transfers between departments or from one shift to another or from a nurse to a doctor). SBAR stands for

- Situation: Describe what's happening. Highlight aspects that need attention.

- Background: Provide history. Put problems in perspective. Highlight trends.
- Assessment: Provide your judgment about the situation.
- Recommendation: Explain what you think the person taking over should do.

Another approach for improvement in health care is the initiatives that the Institute for Healthcare Improvement (IHI) has taken. On December 12, 2006, the IHI announced, with the support of prominent leaders in American health care,[5] a national campaign to dramatically reduce incidents of medical harm in U.S. hospitals. The Five Million Lives Campaign asked hospitals to improve the care they provide in order to protect patients from 5 million incidents of medical harm over a 24-month period, ending December 9, 2008. This represents a continuation of the largest improvement effort undertaken in recent history by the health care industry (Rao and Hoyt 2010).

No one in health care can feel comfortable with the magnitude of infections, adverse drug events, and other complications that hospital patients endure. Dozens of organizations and programs are now working to reduce that toll. They deserve encouragement. This campaign joins those efforts and seeks leverage and scale that our nation has never had before to make care safe everywhere.[6] We can, and should, equip all willing health care providers with the tools they need to make the motto "first, do no harm" a reality.

IHI estimated that 15 million incidents of medical harm[7] occur in U.S. hospitals each year (High Beam Research 2007). This estimate of overall national harm is based on IHI's extensive experience in studying injury rates in hospitals, which reveals that between 40 and 50 incidents of harm occur for every 100 hospital admissions. With 372 million admissions in the United States each year (according to the American Hospital Association's National Hospital Survey for 2005), this equates to approximately 15 million harm events annually or 40,000 incidents of harm in U.S. hospitals every day.[8]

The improvement in health care—as far as the IHI is concerned—especially to achieve patient safety and high reliability in their organizations is based on eight steps. They are as follows:

- Step One: Address strategic priorities, culture, and infrastructure
- Step Two: Engage key stakeholders

- Step Three: Communicate and build awareness
- Step Four: Establish, oversee, and communicate system-level aims
- Step Five: Track/measure performance over time, strengthen analysis
- Step Six: Support staff and patients/families impacted by medical errors
- Step Seven: Align system-wide activities and incentives
- Step Eight: Redesign systems and improve reliability

However, as important as these initiatives are, they will not improve health care unless management articulates and implements these initiatives in their specific health organizations. We believe that success will come to those who recognize and do something about the leadership in their own organizations. Specifically, we believe that improvement in health care is a function of (a) seven leadership leverage points and (b) specific intervention points. The leadership leverage points are as follows:

- Establish and oversee system-level aims for improvement at the highest board and leadership level
- Align system measures, strategy, and projects in a leadership learning system
- Channel leadership attention to system-level improvement
- Get the right team on the bus
- Make the chief financial officer a quality champion
- Engage physicians
- Build improvement capability

Some of the key intervention points are the following:

- Deploy rapid response teams at the first sign of patient decline and before a catastrophic cardiac or respiratory event.
- Deliver reliable, evidence-based care for acute myocardial infarction to prevent deaths from heart attack.
- Prevent adverse drug events by reconciling patient medications at every transition point in care.
- Prevent central line infections by implementing a series of interdependent, scientifically grounded steps.
- Prevent surgical site infections by following a series of steps, including reliable, timely administration of correct perioperative antibiotics.
- Prevent ventilator-associated pneumonia by implementing a series of interdependent, scientifically grounded steps.

Though all of the above sound wonderful and offer great hope for a general improvement in health care, it is important to recognize that the fundamental change that needs to be recognized is that of a *paradigm shift*. Kuhn (1996) brought to widespread attention the notion of paradigm shift—a change in thinking, structures and processes so radical as to make before and after appear to have little, if any, relationship to one another. Morrison (1996) picked up on this concept of sudden transformation, offering a compelling metaphor for contrasting twentieth-century health care (the first curve) with an entirely new, second curve health care paradigm that is now in its infancy.

The key to understanding the paradigm change in any environment, much less in health care, is that the road to *the* change is not linear. Rather, it is quite revolutionary, with many discontinued and evolutionary bumps on the way. In other words, it is a traumatic experience for those involved in the process.

If we look historically at the health care industry—an industry approaching \$2 trillion annually, the largest of our society—we will notice that this huge industry is based on an eighteenth-century model. This means (a) train the craftspersons (physicians, nurses, and so on), (b) license them, (c) supply them with resources, and (d) let them alone as they care for patients. To be sure, this model worked quite well to the point where the health care system in the United States became the envy of the world.

As good as the health care system has been, it has become very complex and fragmented and its infrastructure now borders on being ineffective. Part of this is because we have fallen victim to our own success. Care was defined by individual encounters between patients and clinicians, and costs were both moderate and covered by affordable health insurance. But we are now paying a huge price for having isolated medical and nursing practice from the management of resources. In particular, we must now devote massive creative resources to thoughtfully and comprehensively designing the physical and information infrastructures that are now absolutely essential (Merry 2003).

The road to this change demands conscious redesign of the system, from both its delivery and management stakeholders, including insurance providers. Perhaps more important, the change has to be in our assumptions and beliefs toward a new holistic health system (systems thinking approach). The new system will replace the old assumptions and beliefs but will also increase the efficiency and care for the patient and will control costs to a certain degree.

As surprising as it may sound, the changes are already occurring. A great example was given by Merry (2003). What Merry described is not simply a process. He represents a paradigm shift from a medical doctor's needs to the patient's needs. He represents a thoughtfully conceived and interlinked system by a way of a typical breast cancer diagnosis in a time frame of the 1920s to the present and the Nicollet Medical Center model from 1993 to the present. In the first case the cycle time was 1–8 weeks; in the second, the cycle time was cut to 2 hours. Furthermore, it is a system carefully designed around the patient, with the dual goals of clinical excellence all caregivers seek and maximum ease of use and comfort for patients.

We all want to do something to alleviate this pain, and we are in need of far more collaborative approaches as we pursue second curve systems innovation. Those wanting to change the health care system need to ask the following questions:

- As much as I want to help, does my thinking still trap me into tweaking failed first curve processes?
- Am I able to examine deeply and question my own assumptions and beliefs?
- Is the veil that still shrouds health care's second curve beginning to lift for me?
- Am I willing to explore with others and truly engage in the building of an American health care second curve that might be even more spectacular in its ability to serve all stakeholders than the first?

Improvements have been made, no doubt about it. We see the Sisters of St. Mary Health System in St. Louis boasting a Malcolm Baldrige National Quality Award winner; we see the Henry Ford Hospital System in Detroit incorporating Six Sigma methodology and many other medical institutions with tremendous innovations and improvements. However, the industry as a whole is only beginning its journey toward institutionalizing Six Sigma or any other quality initiative that is near perfection for such important measures as those relating to patient harm.

Merry (2003) identified some of the assumptions and beliefs for both the pre- and postindustrial age, which he addressed as curves. (The notion of the curve is based on the S-curve. The S-curve is a well-known project management tool and it consists in "a display of cumulative costs, labor hours or other quantities plotted against time (Wideman 2002)." The name derives from the S-like shape of the curve, flatter at the beginning and end and steeper in the middle, because this is what most of the projects look like. As time goes on, the curve shifts upwards because of changes in the project.) Merry's (2003) list is shown here.

- The preindustrial (first curve) age:
    - Believes quality capability of 2 to 4 sigma is satisfactory.
    - Organized around needs of providers.
    - Asks the community to come to provider.
    - Reacts only to the individual (a diabetic, for example), relatively blind to needs of population (diabetics, for example).
    - Providers define quality in terms of morbidity and mortality and resist publication of actual data because "our patients are different."
    - Conceives quality capability almost solely in terms of professional skills, with virtual blindness to importance of support systems.
    - Assumption: "First, do no harm." Provider intentions impeccable.
    - Reality: Human error generates harm, with threat of punishment as a deterrent.
    - Complexity makes it easy to do things wrong, hard to do things right (Kohn, Corrigan, and Donaldson 1999).
    - Solution to problems translates to retraining or censuring professionals and provider institutions.
    - Ultimate definition of quality endlessly debated, thus avoiding adequate measurement, management, and improvement.
    - System is fragmented. Patient fends for her or himself, moving from silo to silo.
    - Medical record is fragmented and idiosyncratic to particular silo. Individual caregivers work off entirely unconnected, often contradictory scripts.
    - Information is centralized and hierarchical. Physician is supreme source of knowledge and dictator of therapy.
    - Insurance is monolithic, not enrollee sensitive, with perhaps a few choices for individuals.
    - Billing and payment systems are arcane, confusing, and virtually impossible to understand.
    - Payment system is blind to quality and value and rewards volume, even that generated by poor quality and error.
    - Huge resources are consumed in reimbursing inefficient systems, human error, litigation, and cost-plus models.

- Health care is an isolated, quirky, high-tech, organizationally primitive industry, a throwback to pre-eighteenth-century human organizational development.
- Crashes are common, and medical error death and injury headlines are regular, predictable occurrences.
- As of 2003, trust in the system is increasingly shaky and falling.
- Extremely high, probably incalculable cost of poor quality exists.
- The postindustrial (second curve) age:
  - Believes quality capability of 5 to 6 sigma is essential.
  - Designed around needs of those served, including those of all caregivers.
  - Providers reach out to where community lives.
  - Plans for population reduces need for individual care but retains ability to respond to needs of the individual.
  - Users add to definition of quality, including satisfaction with service, functionality, and value; insist on information to choose, using appropriately case mix–adjusted information.
  - Understands that carefully designed quality infrastructure is absolutely essential to reduce risk and optimize skills of professionals.
  - Assumption: Humans are inherently fallible, and harm occurs despite providers' best intentions.
  - Reality: System is error tolerant, accepting human error as inevitable. Designs error proofing.
  - Well-designed latent workplace conditions make it easy to do things right and hard to do things wrong.
  - Solution to problems translates to redesigning systems to become more error tolerant and human supportive.
  - Consensus exists regarding a variety of key measures, including access to care, clinical outcomes, functionality, satisfaction, and value received.
  - System is seamless. Coordinates needs of complex patients, using case managers for those that are especially difficult.
  - Medical record is electronic and instantly updated and available for all relevant caregivers; all caregivers read from precisely the same script.
  - Information is dispersed. All caregivers and patients have direct access. Physician is integrator and facilitator of choices.

- Insurance is mass customized. Web-based options are chosen by individuals based on specific needs.
- Coverage and copayments are clear, web facilitated, and easy to navigate.
- Payment system is fine-tuned to value and rewards superior performance as defined by value equation (Value = Quality/$).
- Huge resources are freed up for innovation and quality improvement, with cost-plus, value-blind reimbursement a distant memory.
- Health care is a vibrant participant in the best that learning from the industrial and information revolution can offer.
- Crashes are rare, with medical error death equivalent to airline performance.
- As of 2000, trust in the system is high and rising.
- Minimal cost of poor quality exists.

How do we expedite this paradigm shift? By providing to the medical profession at large both appropriate education and training in at least the following areas:

- *Ethics and communication*: At the outset of the Hippocratic Oath, ethics and communication emerged by default. Medical professionals, whether they like it or not, cannot escape ethical dilemmas. Issues about truth-telling, confidentiality, informed consent, end-of-life care, allocation of scarce resources, religious considerations, and many other issues are all part of their domain.
- *Information technology*: As observed by many professionals within and outside health care, the information technology in a wide range of areas (e.g., including medical records, drug ordering, and many more) is the slowest sector.
- *Management*: Doctors are leaders of as well as participants in medical, research, or administrative teams. They need to know how to account for resources and to plan strategically. They need training in negotiation skills, evaluation performance in both patient and organizational changes, and certainly improvement of patient care. All of these are issues of management, and they can be learned.
- *Statistics*: Sample size, odds ratios, specificity, variability, and predictive values are essential for the modern medical professional. Yet, statistics education is not required in medical school.

- *Law*: Medical professionals are told what they can and cannot do via the law and court rulings. It is imperative to recognize that issues dealing with life and death, communicable diseases, stem research, malpractice, and other ethical issues provide a gap in their training and education and need to be fixed.
- *Medical economics*: At a time when health care consumes over 15 percent of our domestic product, health economics can no longer be considered peripheral to medical education.

This education should be undertaken in both premed and medical schools, as well as a lifelong training. In the formal education process, these students, depending where they are (undergraduate or graduate school), should be able to take these disciplines from the business school, college of liberal arts, or engineering school. The focus should be on creating leaders that have points of view on topics that are vital to their success and the success of the organizations they lead. We are not suggesting that medical professionals do not have their own points of view. Rather, we emphasize that these medical leaders need to transfer their knowledge to others and always have leadership teachable points of view that are based on the most recent research and practices. Furthermore, they must be able to challenge their own beliefs and paradigms, for example, in the areas of:

- How to grow a business
- How to exceed customer expectations
- How to achieve one's potential
- How values and ethics are essential for success
- How teaching, coaching, or mentoring brings out the best in others
- How to use courage and fierce resolve to achieve goals and work with others
- How humility plays a role in successes and effectiveness

FINANCIAL

In the last 10 years the financial world has experienced some very serious hits in their profitability as a result of legal difficulties. Many banks have suffered because of lack of internal controls in companies like Enron, WorldCom, and many others. Some examples of how much banks had to

TABLE 1.3

Penalties Due to Financial Irregularities ($ in millions)

Citigroup	$2,600	Lehman Bros.	$63
JP Morgan	$2,000	BNP Paribas Securities	$38
Bank of America	$461	Caboto Holding	$38
Deutsche Bank	$325	Mizuho International	$38
ABN Bank	$278	Credit Suisse First Boston	$13
Mitsubishi Bank	$75	Goldman Sachs	$13
WestLB	$75	UBS Warburg	$13

pay in fines due to financial irregularities and reported by the Office of New York State Comptroller are summarized by Davis and Norris (nd) and are shown in Table 1.3.

The scandals were so great that the U.S. government decided to pass legislation so that failures of that magnitude would not happen. That legislation is known as the Sarbanes–Oxley Act of 2002. It states that it is "an act to protect investors by improving the accuracy and reliability of corporate disclosures made pursuant to the securities laws, and for other purposes (Section 404)."

The Sarbanes–Oxley Act was designed to restore investor confidence after a string of corporate accounting frauds.[9] The law is changing corporations as profoundly as securities laws did after the crash of 1929. It has 11 parts:

- Title I. Sets up public company accounting oversight board, to oversee public company auditors.
- Title II. Sets rules to ensure that auditors are independent and not biased against investors.
- Title III. Strengthens rules about financial reporting, auditing, and corporate governance.
- Title IV. Increases corporate financial disclosure.
- Title V. Sets rules requiring unbiased research from securities analysts.
- Title VI. Gives Securities and Exchange Commission and federal courts power to enforce the new rules.
- Title VII. Calls for studies of firms that provide accounting and financial services.
- Title VIII. Sets punishments for corporations guilty of accounting fraud.

- Title IX. Adds to penalties for individuals guilty of corporate fraud.
- Title X. Recommends that public company tax returns be signed by CEOs.
- Title XI. Stiffens penalties for record tampering or other hindering of corporate fraud investigations.

However, after its enactment, both the Bush administration and many business interests were arguing that the time was right to loosen some of the requirements of the Sarbanes–Oxley corporate reform law, passed after the scandals at Enron and WorldCom. But if anyone doubts that these reforms, designed to increase accuracy and accountability, were necessary, consider this: According to the research firm Glass Lewis (Lucich 2007), nearly 10 percent of companies listed on exchanges in the United States refiled their 2006 financial statements after finding mistakes.

In those cases investors were making decisions based on incorrect information, and some executives were being paid for results they did not achieve. These are often more than small bookkeeping errors. For example the week of March 11, 2007, General Motors restated 5 years of financial results. In its annual report, the company warned that the lack of effective internal controls "could adversely affect our financial condition and ability to carry out our strategic business plan (GM Annual Report 2007, p. 3)."

Internal controls are the methods that companies use to ensure that their financial statements are accurate, like reconciling cash on a company's books with its actual bank statements or running built-in software checks of accounts. They include such simple steps as having a code of ethics and determining whether the company has sufficient accounting staff.

Sarbanes–Oxley's Section 404—much maligned by businesses as too expensive and onerous—requires that these controls be tested by management and checked by outside auditors. Sometimes problems crop up where one might least expect them.

Wehner (2005) summarizing the Glass Lewis report notes that management at Qwest Communications manipulated the account of accrued employee vacation days as part of its effort to make the company look healthier than it was.

There is a silver lining in the report, one that supports Sarbanes–Oxley. Regulators have delayed imposing Section 404 compliance on smaller companies. But among the larger companies that have had to comply, such restatements fell 14 percent in 2006. Though the new rules may be a hassle, it is clear that auditing internal controls is helping companies clean up their acts, and it would be a serious mistake to weaken Sarbanes–Oxley.

As optimistic the Sarbanes–Oxley Act continues to be, the fact is that it is not the answer. In the fall of 2008, we saw financial institutions go out of business because of manipulating the risk with investments and faulty data. Some of the results for such manipulations were Washington Mutual (an alleged Six Sigma company) and Wachovia Bank. Other companies with troubles were Merrill Lynch, Citigroup, Morgan Stanley, Bear Stearns, Lehman Brothers, UBS, Goldman Sachs, and AIG, not to mention the devastating lows of the stock market.

Other than the bankrupt companies, the bailout continues for major companies. For example, Merrill Lynch, $6.2 billion by Singapore's Temask Holdings and Davis Selected Advisors; Citigroup, $7.5 billion by Abu Dhabi Investment Authority and an additional $75 billion by the U.S. government; Morgan Stanley, $5 billion by China's sovereign-wealth fund; Bear Stearns, $1 billion each by U.S. investor J. Lewis and China's CITIC Securities; UBS, $9.8 billion by the government of Singapore Investment Corporation and $1.8 billion by unnamed Middle East investor (believed to be either Abu Dhabi or Oman entities; Hahn 2008); AIG, $150 billion by the U.S. government.

It seems that the Financial Accounting Standards Board (FASB) is on another one of its intellectually stimulating but oh-so-impractical tangents in testing its new format for accuracy in financial statements. Before companies are forced to pursue this expensive exercise in futility, FASB needs to remember Rule #1 for financial statements: their primary intent is to provide timely, accurate, and useful information to users.

FASB needs to ask users what they find really useful in financial statements, rather than what strikes accountants' fancy. Users and providers of financial statements would be far better served if FASB spent its time defining *earnings* before *interest, taxes, depreciation,* and *amortization* (EBITDA). Every fundamental analysis of a business starts or ends with this metric. But FASB and the SEC seem to suffer from NIH ("not invented here") syndrome. Again, they need to get their heads out of the theoretical sky and put their feet on the ground where financial statement users—that is, their constituency—have to operate.

Until recently, the health of banks has not been a pressing issue for finance executives. After all, it had been almost two decades since the last swell of commercial bank failures, 10 years since a financial crisis of any consequence, and 7 years since the credit cycle took a big dip. Along the way, the earnings of commercial banks quadrupled. But thanks to the subprime crisis, banks and financial institutions are back on everyone's radar. The discovery of modern banking's soft underbelly has created unease. And the ripple effects are hitting corporations from many directions.

Now that the gaping holes in their risk management have been revealed, banks are trying to restore faith. Giants such as Merrill Lynch and Citigroup are really changing their risk structures or simply throwing new C-suite executives at the problem. One important question: How involved will bank CFOs be in monitoring exposures in financial markets?

Even as banks work out their risk quandaries, they are already demonstrating far less appetite for corporate loans, not to mention how to find capital for new investment—and paying more for it.[10]

The clots in the arteries of financial markets are also bad news for companies that grant trade credit to their customers. The National Association of Credit Management's Credit Manager's Index, an indicator of trade credit trends and receivables' performance, has dropped for six straight months. Companies could be paying for the mistakes of banks for years. But in times of crisis, much can be learned. The credit crunch of 2007 and the collapse of some of the biggest financial institutions in the United States and the world will make CFOs more inquisitive of their financial institutions' business practices and balance sheets. That may do more to strengthen the foundations of banking than the previous 20 years of prosperity (Ryan 2008; see note 7).

In financial markets, as soon as you hear the words new paradigm you know that the next catastrophe is not far away. The reasons are not complex, nor are the observations opaque. It does not matter whether it is dotcoms or subprime mortgages. Sooner or later in market cycles the main participants, and especially the banks, become carried away with a Wildean[11] belief in their own, oft-declared genius (Morley 2008).

Morley continues by saying that the gains are magnified by leverage and the egos inflate in direct correlation to the paper profits. At this point, the Darwinian principle that he who makes the profit must be both protected from the rules applied to the rest and rewarded irrespective of risk start to apply. In the recent case of subprime mortgages, bankers lent money to anyone, regardless of their credit history, because the risk was securitized and the financial equivalent of explosive pass-the-parcel ensued. When the game ended, bankers walked away with much of the gains, and the parcel exploded in all our faces. The paper gains disappeared and the losses were added to our tax bill.[12]

This is real moral hazard. It just happens more quickly in bear markets. In the midst of this, the regulators tinker with the safety rules of the *Titanic* (it always sinks, regardless). What they cannot change is greed and stupidity. *Moral hazard* is when risk and reward have an asymmetrical relationship—usually a lot of reward for one person and most of the risk for the other. This is the root cause of the subprime and credit crisis. And it is at the heart of most financial meltdowns. They just manifest themselves differently and therefore catch us unawares. It is like generals planning for the next war based on the experience of the last one; and just as failed generals get medals, bankers get bailed out (Morley 2008).

Whereas it has been customary to reward success, now it seems that failure is rewarded even more so than success. In fact, the reward is substantial and guaranteed by the government. The moral here is that there are no consequences and no accountability, not to mention responsibility for any bad decisions. When an Enron, or even a Barings, fails in isolation and there is no general market failure, the regulators are sanctimonious about the dangers of systemic moral hazard, because collateral market damage is not perceived as a great risk. But when a Northern Rock or Bear Stearns is about to fail in the midst of a market crisis, they have to be bailed out regardless of moral hazard. Until we can think of another source of mass access to paper and electronic debt that we call "money," every time the banks mess it up they will always be bailed out. This is a problem not just for banking; it is a moral question for society. After all, where do we stop (Morley 2008)?

Do not look to politicians for solutions. There are too many examples of egregious moral hazard in that group. But if banks are to be supported with generous liquidity or guarantees underwritten by the taxpayers—or even recapitalized with taxpayers' money—they should use part of the money to try to prevent defaults on mortgages by all but their most delinquent customers. Short-term support from the public authorities may prevent a complete meltdown, but in the long term it creates dependency. If the banks do not behave properly as the credit crisis turns to economic grief—the precedents are not good—then when markets return to normal it may be necessary to take the cheap loan benefits back in special taxes.

There are also things the banking sector can do to help itself. Strong, independent, nonexecutive directors must control remuneration so that talented executives are properly rewarded in a competitive market but not by allowing them to bet the banks' and shareholders' money while themselves sharing only in gains but not losses. If executives share the risks as well as the rewards, moral hazard reduces in proportion.

Times like these are cathartic. Bubbles of overpriced assets collapse along with the egos of many investors. The wannabe stars in, say, hedge funds and private equity will go to the wall, but the genuinely talented will survive. Some senior banking heads have rolled—and more will no doubt need to roll—but the danger is that the banks learn nothing, only to repeat it all in a few years' time (Morley 2008).

Finally, a word of caution as far as "scoring scorecards" is concerned. Data companies and financial professionals (including accountants) love all data and its manifestations. The more data they have, the better they feel. Of course, data requires collection, storage, and parsing. However, in the labyrinth of the data, they ignore two key facts:

- Too many measurements often lead to conflicts, and there is no mechanism other than managerial judgment to resolve the conflicts in scorecards.
- There are no successful scorecards that tie into value programs, because the time horizons for scorecards tend to be daily, weekly, or monthly, and value is the present value of future cash flows; hence a structural conflict in the application.

In addition to these primary flaws, scorecards are almost always static, and markets and circumstances change faster than scorecards do. In fact, all of that useless data is itself an impediment to value-creating behavior, because it tends to make people try to game the system.

This is not to say that measurement is not important; it is vital, but scorecards are only a small part of the answer in operations and worse than useless in decisions about strategy.

SUMMARY

In this chapter we discussed the need for quality improvement by examining six diverse sectors of our economy. In the next chapter we address the issue of leadership and its importance in the organization that is about to embark on quality initiatives.

ENDNOTES

1. Portions of this section have been adopted from Hill (2006).
2. More info on the CEP may be found on their website: http://www.character.org.
3. *Ring fencing* is separating something from usual judgment and guaranteeing its protection, especially the funds of a project. Generally it is associated with Government health plans such as the European (the United Kingdom's, France's and Germany's) plans. Specifically, this term has been in use since the 1980s to denote the funds that are set aside for a project and cannot be spent on anything else. (source: http://www.phrases.org.uk/meanings/302450.html)
4. A *wonk* is a person preoccupied with arcane details or procedures in a specialized field.
5. The campaign was able to focus on the following steps to reduce harm and deaths:
   - Deploy Rapid Response Teams ... at the first sign of patient decline
   - Deliver Reliable, Evidence-Based Care for Acute Myocardial Infarction ... to prevent deaths from heart attack
   - Prevent Adverse Drug Events (ADEs) ... by implementing medication reconciliation
   - Prevent Central Line Infections ... by implementing a series of interdependent, scientifically grounded steps called the "Central Line Bundle"
   - Prevent Surgical Site Infections ... by reliably delivering the correct preoperative antibiotics at the proper time
   - Prevent Ventilator-Associated Pneumonia ... by implementing a series of interdependent, scientifically grounded steps including the "Ventilator Bundle"
6. Several preeminent authorities on patient safety and health care improvement have reviewed and endorsed this estimate and the campaign's aim. These include Lucian Leape, M.D., Adjunct Professor of Health Policy at the Harvard School of Public Health; Brent James, M.D., Executive Director of Intermountain Healthcare's Institute for Healthcare Delivery Research; Ross Baker, Ph.D., Professor of Health Policy, Management and Evaluation at the University of Toronto; and David Bates, M.D., Medical Director of Clinical and Quality Analysis, Information Systems at Partners HealthCare System.
7. IHI defines *medical harm* as unintended physical injury resulting from or contributed to by medical care (including the absence of indicated medical treatment) that requires additional monitoring, treatment, or hospitalization.
8. Based on the results of a survey the IHI has recommended new interventions targeted at harm. They are as follows:
   - Prevent methicillin-resistant *Staphylococcus aureus* (MRSA) infection by reliably implementing scientifically proven infection control practices throughout the hospital.
   - Reduce harm from high-alert medications starting with a focus on anticoagulants, sedatives, narcotics, and insulin.
   - Reduce surgical complications by reliably implementing the changes in care recommended by the Surgical Care Improvement Project (SCIP).
   - Prevent pressure ulcers by reliably using science-based guidelines for prevention of this serious and common complication.
   - Deliver reliable, evidence-based care for congestive heart failure to reduce readmissions.

- Get boards on board by defining and spreading new and leveraged processes for hospital boards of directors, so that they can become far more effective in accelerating the improvement of care.

9. Whereas the Sarbanes–Oxley Act legislation is the law of the land in the United States, in Europe and in the rest of the world, assurances about financial integrity are provided with the international financial reporting standards (IFRS). As of this writing, the discussion is heavy and there are still strong opinions regarding which of the two is better and more effective. What is certain is the fact that there is movement toward the IFRS. The currently proposed timetable would allow the largest U.S. companies to begin reporting their 2009 results in IFRS and then require large accelerated filers to begin using IFRS in 2014, followed by other accelerated filers in 2015 and smaller companies in 2016. Companies would have to provide financial statements from the previous 3 years in IFRS at those times as well for comparability. Financial Accounting Standards Board (FASB) and the IASB have already committed to converging their respective standards by 2011, which, if accomplished, would effectively blur the distinctions between the two.

   It must be noted here that the IFRS is less robust compared with generally accepted accounting principles (GAAP) in the United States. Also preliminary research from Europe shows that the international standards in fact afford investors little comparability among financial statements, one of the key reasons given for U.S. convergence. Furthermore, the possibility of letting the International Accounting Standards Board (IASB) be the standards-setter for the world, given its recent capitulation to pressure from European Union authorities to loosen fair value accounting for banks, presents a problem for many U.S. specialists and financial institutions, including the Securities and Exchange Commission (SEC).

   In addition to the Sarbanes–Oxley Act and the IFRS standards, the promise or threat, depending on your perspective, of a mandate to file financial statements in a new "interactive data" format has been looming for a long time. However, the time has arrived when the SEC seems poised to announce mandatory filings of eXtensible Business Reporting Language (XBRL) as a means of vastly improving transparency, data quality, and efficiency for investors, both institutional and individual. Rather than waiting for the inevitable, many public companies are taking advantage of the SEC's Voluntary Filing Program (VFP) to learn what XBRL is all about, how it will impact their SEC filings, and whether XBRL can provide them with additional benefits.

10. One wonders, even with the mess that the financial institutions find themselves in, why they continue to have excess in spending even after the billions of infused dollars by the U.S. government. Instead of availing the money to organizations that needed it and offering reasonable loans, they have chosen to buy more banks and hold the economy at risk.

11. The power to give an entirely new interpretation to ideas.

12. The subprime fiasco started under the Carter administration, when the government under the Community Reinvestment Act (CRA) forced the financial institutions to give loans to people with low income and little or no credit for a mortgage whether they were qualified or not. This was a catastrophic—as it turned out—mess of the government's attempt to engineer housing outcomes. The rationale for such a deal was the notion that *home ownership* is a right and therefore everyone must have a home. It took about 30 years, but it worked as over 70 percent of the population became homeowners. However, in the process, the financial institutions suffered and in some cases went out of business.

REFERENCES

American Hospital Association. http://www.aha.org/aha/resource-center/index.html

Chantapalaboon, S. (2008). "A unique marketing weapon." Available at: www.nationmulti-media.com/2008/08/27/.../business_30081613.

Chao, L. (2008). "Drawbacks of climate surveys." *The Nation* 1 September.

Davis, F. and J. Norris. (nd). http://www.davisnorris.com/securities_fraud.shtml

Ellis, B. (July 26, 2008). *Dakota Voice.* http://www.dakotavoice.com/2008/07/only-a-virtuous-people-are-capable-of-freedom/

Enterpreneur International Magazine. (2007). http://www.entrepreneur.com/franchises/toplists/2007.html

Erixon, F. (June 12, 2008). Europe should not resist free trade in health care. http://www.ft.com/intl/cms/s/0/d713b738-3895-11dd-8aed-0000779fd2ac.html#axzz1SkSMES4p

Gitlow, H. S. and S. J. Gitlow. (1987). *The Deming guide to quality and competitive position.* Englewood Cliffs, NJ: Prentice-Hall.

GM. 2007 annual report

Hahn, A. L. (2008). "How poor risk management techniques contributed to the subprime mess." *CFO* March, 51–58.

Hall, E. (1959). *The Silent Language.* New York. Doubleday

High Beam Research (January 1, 2007) http://www.highbeam.com/doc/1G1-158090795.html

Hill, D. (October 2006). What lies beneath. Quirk. p. 68. http://www.quirks.com/articles/2006/20061006.aspx?searchID=185734356&sort=9

Indian Express. http://www.indianexpress.com/news/just-what-the-hospital-ordered-global-accreditations/12890/

Institute of Medicine. (1999). http://md-jd.info/abstract/Institute-of-Medicine-Report.html

Jones, B. (May 17, 2007. Big Three's woes extend beyond high labor costs. USA Today. p. B2.

Kohn, L. T., J. M. Corrigan, and M. S. Donaldson, editors. (1999). *Committee on Quality of Health Care in America, Institute of Medicine.* Washington, DC: National Academies Press.

Kuhn, T. (1996). *The structure of scientific revolutions,* 3rd ed. Chicago: University of Chicago Press.

New York: Simon & Schuster.

LeDoux, J. (1996). *The Emotional Brain: The Mysterious Underpinnings of Emotional Life.* McKinzey, A. http://en.wikipedia.org/wiki/Medical_tourism_in_India

Lucich, T. (March 20, 2007). Keep Sarbanes-Oxley. http://insidesarbanesoxley.com/2007/03/keep-sarbanes-oxley/

Merrill Lynch. (August 2005). Global Private Client Investment Management and Guidance. Asset Class Assumptions. A resource Guide to Merrill Lynch Financial Advisors. NY. NY.

Merry, M. D. (2003). "Healthcare's need for revolutionary change." *Quality Progress* September, 31–35.

Morley, I. (June 12, 2008). When you hear 'new paradigm' head for the hills. http://www.ft.com/intl/cms/s/0/e8e7e73c-3894-11dd-8aed-0000779fd2ac.html#axzz1SkSMES4p

Morrison, I. (1996). *The second curve: Managing the velocity of change.* Darby, PA: Diane Publishing.

Newmark, A. (1999). http://philadelphialawyer.info/Institute-of-Medicine-Report.html

Novation Group. (November 2007). Survey: Employers Plan to Raise Diversity Training Spending This Year. http://clomedia.com/articles/view/survey_employers_plan_to_raise_diversity_training_spending_this_year

Program for International Student Assessment Scores. (2009). http://online.wsj.com/public/resources/documents/st_PISA1206_20101207.html

Rao, H. and Hoyt, D. W. (March 5, 2010). Institute for Healthcare Improvement: The 5 Million Lives Campaign. http://hbr.org/product/institute-for-healthcare-improvement-the-5-million/an/L16-PDF-ENG

Rhodes, F. H. T. (2006). "Sustainability: The ultimate liberal goal." *The Chronicle of Higher Education* 20 October.

Ryan, V. (2008). "Companies in all industries are paying for the transgressions of the banking sector." *CFO* March, 49.

Sarbanes–Oxley Act of 2002. (Pub.L. 107-204, 116 Stat. 745, enacted July 30, 2002).

Shunk, C. (November 11, 2010). Report: Automaker R&D spending took huge hit during recession.http://www.autoblog.com/2010/11/11/report-automaker-randd-spending-took-huge-hit-during-recession/.

Wehner, R. (May 17, 2005). Investor reports blast Qwest board, exec pay . *The Denver Post.* http://www.denverpost.com/business/ci_2740474#ixzz1RdtSm7Jx ; http://www.denverpost.com/termsofuse

Wideman, R. M. (March 2002). Wideman Comparative Glossary of Common Project Management Terms, V2.1. Retrieved from http://www.maxwideman.com/pmglossary/

Zaltman, G. (2003). *How Customers Think: Essential Insights into the Mind of the Market.* Boston: Harvard Business Press.

SELECTED BIBLIOGRAPHY

Bernstein, J. (2006). *All together now: Common sense for a fair economy.* San Francisco: Berrett-Koehler Publishers.

Boring, E. G. (1923). "Intelligence as the intelligence tests see it." *The New Republic* June, 35–37.

Brown, T. (1995). "Great leaders need great followers." *Industry Weekly* 244(16): 25.

Burbules, N. C. (1990). "The tragic sense of education." *Teachers College Record 91,* 469–479.

Burbules, N. C. (2000). "Aporias, webs, and passages: Doubt as an opportunity to learn." *Curriculum Inquiry 30,* 171–187.

Burbules, N. C. (2004). "Ways of thinking about educational quality." *Educational Researcher* August–September, 4–10.

Burbules, N. C. and R. Berk. (1999). "Critical thinking and critical pedagogy: Relations, differences, and limits." In *Critical theories in education,* edited by T. S. Popkewitz and L. Fendler. New York: Routledge.

Chaleff, I. (1995). *The courageous follower.* San Francisco: Berrett-Koehler Publishers.

Crago, M. C. (2000). "Patient safety, Six Sigma and ISO 9000 quality management." *Quality Digest* November, 37–40.

Crockett, W. J. (1981). "Dynamic subordinancy." *Training and Development Journal 35* (May 1981), 155–164.

Derrida, J. (1981). *Plato's pharmacy, dissemination*. Chicago: University of Chicago Press.

Dewey, J. (1899/1980). *School and society*. Carbondale: Southern Illinois University Press.

Dewey, J. (1916). *Democracy and education*. New York: Macmillan.

Dixon, G. and J. Westbrook. (2003). "Followers revealed." *Engineering Management Journal* 15(1): 19–25.

Epstein, R. A. (2006). *Overdose: How excessive government regulation stifles pharmaceutical innovation*. New Haven, CT: Yale University Press.

Erixon, F. (2008). "Europe should not resist free trade in healthcare." *Financial Times* 12 June.

Frisina, M. (2005). "Learn to lead by following." *Nursing Management* 36(3): 12.

Gerlin, A. (2000). "How your hospital could kill you." *Readers Digest* June, 174.

Gould, E. (2007). "The health–finance debate reaches a fever pitch." *The Chronicle of Higher Education* 13 April.

Gratzer, D. (2006). *The cure: How capitalism can save American health care*. New York: Encounter Books.

Hacker, J. S. (2006). *The great risk shift: The assault on American jobs, families, health care, and retirement and how you can fight back*. New York: Oxford University Press.

Hirsch, E. D. (1988). *Cultural literacy: What every American needs to know*. New York: Vintage.

Kelley, R. (1992). *The power of followership: How to create leaders people want to follow and followers who lead themselves*. New York: Doubleday Currency.

Kling, A. (2006). *Crisis of abundance: Rethinking how we pay for health care*. Washington, DC: Cato Institute.

Lundin, S. C. and L. C. Lancaster. (1990). "Beyond leadership: The importance of followership." *The Futurist* May–June, 18–22.

Nolan, J. S. and H. F. Harty. (2001). "Followership > leadership." *Education* 104(3): 311–312.

Noris, S. P. (1998). "Intellectual independence for nonscientists and other content-transcendent goals of science education." *Science Education* December, 239–258.

Nussbaum, M. C. (1997). *Cultivating humanity: A classical defense of reform in liberal education*. Cambridge, MA: Harvard University Press.

Ohsfeldt, R. L. and J. E. Schneider. (2006). *The business of health: The role of competition, markets, and regulation*. Washington, DC: AEI Press.

Peters, R. S. (1965). "Education as initiation." In *Philosophical analysis and education*, edited by R. D. Archambault. London: Routledge and Kegan Paul.

Pirsig, R. (1974). *Zen and the art of motorcycle maintenance*. New York: William Morrow.

Polanyi, M. (1958). *Personal knowledge*. London: Routledge.

Polanyi, M. (1967). *The tacit dimension*. New York: Anchor Books.

Rancière. M. (1991). *The ignorant schoolmaster*. Stanford, CA: Stanford University Press.

Ryle, G. (1962). *The concept of mind*. New York: Barnes and Noble.

Siegel, H. (1988). *Educating reason: Rationality, critical thinking, and education*. London: Routledge and Kegan Paul.

2

Leadership

In the last chapter we focused on six sectors of our economy to show the need for quality improvement. In this chapter we will address the importance of leadership for any organization that is about to undertake quality improvement.

There is a difference between *leadership* and *management*. Leadership deals with the need for change in an organization, whereas management copes with the procedures and practices of the company. The early part of the twentieth century foresaw the rise of management. The amount of change was small at any period of time and management principles were ideal; that is, the administration of the practices and policies of the company. Leadership was more valuable in the last part of the century and will be needed more for the early part of the twenty-first century. The ability of a manager to lead is important, but it is also important for there to be a balance with administrative or management skills. To paraphrase Kotter (1990), a firm may suffer under strong leadership if it has weak management skills.

Therefore, before we discuss leadership, let us review some of the issues that deal with the functions of management. Typical functions related to management are the following:

- *Planning:* The setting of objectives and means to achieve them. Among the functions of management, planning takes the greatest amount of a manager's time. Planning involves defining the vision, mission, directions, goals, and objectives for the organization. This also includes thinking, gathering of data, analysis of data, creating innovative approaches to strategies, and decision making.
- *Leading:* The influencing and motivating of workers to work toward the goals. Kotter (1982) defined *leading* as *directing*. The

manager accomplishes work through people. The employees must be motivated in some method to perform their tasks. Employees' enthusiasm is aroused to help support the goals and objectives of the organization. Knowledge of human behavior, motivational theory, and communications is required for a manager to effectively lead a department.

- *Controlling (Monitoring):* The process of determining how well objectives are being achieved. The manager needs to be aware of the performance of his or her work unit in terms of actual performance to either budgeted or standard performance. The nature or extent of the gap will provide an indication to the manager as to the corrective action that needs to be taken. The manager with responsibility for controlling and monitoring must follow work progress and communicate regularly with the people. Monitoring of staff activities is done using several methods, including (a) review of memos, reports, and letters from subordinates; (b) meetings, voicemail, and casual conversations with the employees; (c) input from other employees, managers, and clients; and (d) observation of employees while they work.
- *Organizing:* The use of resources such as people, tasks, or machines to do the work. The organizing portion of management requires a structure for the organization. This will usually be viewed as the organization chart. The tasks and work to be performed must be defined, with the proper relationships between jobs and people developed. The proper amount of technology and other supporting resources must be determined for effective operation of the organization.
- *Staffing:* Human resources are part of the people process. The human resources part of the organization usually performs the staffing responsibilities, which include (a) selecting, (b) training and development, (c) appraising, and (d) rewarding of personnel. Of course, in smaller organizations, the supervisor or owner may be performing multiple duties, which include the staffing function.

The practice of management has been in existence for thousands of years. In fact, Xenophon, a Greek mercenary general, wrote the first book on management called *Economy* (Strauss 1970). He was the first to recognize that there is no essential difference between leading soldiers in battle and leading civilian workers on an estate. In other words, for the first time in history we find an author who wrote about leadership as a transferable principle. Xenophon, along with Socrates, did not stop there. They identified several principles and/or characteristics that hold even today. For example, they were aware that some people are more gifted with leadership ability than others, especially when it comes to inspiring others.

Some important management activities include:

- They spend a lot of time with others and at all levels of management.
- The conversations between general managers and others will cover business subjects, and they will ask a lot of questions, with humor thrown in.
- During the conversations, few "big" decisions will be made, and few orders are given, but general managers will do a lot of influencing.
- Though there may be planned meetings during the day, a typical day includes many unplanned items.
- Many of the discussions are short and unplanned.
- The typical general manager works about 60 hours per week.

It is important to recognize that there are different management styles (Council of Indiana 2000) and, depending on the organization (function and size), these styles will influence the behavior of the organization as well as the competition. Typical styles, according to the Council of Indiana (2000) and Galbraith (1995), include the following:

- *Systematic Management*: This was one of the first approaches to the management of the business. The organization of the late 1880s was growing and needed a method of operation. This management style used procedures for control of the company. It had strong clarification of duties and responsibilities with systematic techniques and processes. The emphasis was on economical operations and cost control. The workers accepted, perhaps reluctantly, the strict control of their lives by the company when they were in the workplace.
- *Scientific Management*: Frederick Taylor is the man associated with this management method. He developed and used scientific methods to analyze the workplace and to find the "one best way" for managing the shop. Trained engineers would break down the tasks and find the best way to perform the task. The pig iron worker "Schmidt," a "high-priced man," or a "first-class man" was Taylor's subject in his study of proper tools and techniques. This was the beginning of the industrial engineering approach to management. The scientific management technique called for the following:

- The use of scientific methods (time study, motion study, etc.) to determine specific work elements.
- The proper selection and training of workers, using the "right" methods. Workers doubled or tripled their output and increased their wages by 60 percent.
- The concepts of organization and management, the breaking down of tasks into manageable elements, with specialized staff help. Management does the thinking, and workers do the muscle work.

  Henry Ford built his Model T using the scientific management method and was quite successful. Taylor's (1911) classic text, *The Principles of Scientific Management*, was a dominant textbook used by management in the early part of the century. The thoughts and concepts of the Taylor system are still very much in use throughout this country. Deming's total quality principles are an extension of these principles. Delavigne and Robertson (1994) detailed the differences between the Taylor and Deming philosophies as shown in Table 2.1.

- *Administrative Management*: Administrative management was the beginning of the emphasis on management being a profession that can be learned. There are basic management principles that can be defined, learned, and applied throughout most industries. This would imply that a good manager can be successful in one industry

TABLE 2.1

Comparison of the Taylor and Deming Philosophies

Taylor Method	Deming Method
Professional management controls the business	Use leadership and cooperation
Management makes improvements happen through division of labor	Improvements occur through division of labor, use of creativity and information
Develop systems to perform repetitive tasks	Develop systems to perform repetitive tasks
"One best way" can be created	There is no one best way; variation constantly occurs in a system
With a properly designed system in place, only outside influences can affect the system	There is a need to understand the variation affecting a system
Management must control the variables of the system: workers, selection, motivation, training, supervision, etc.	Create a secure environment, without fear, to improve the system

and, by using his management skills, could switch to another industry with equal success.

- *Human Relations*: The famous Hawthorne studies of the 1920s and 1930s had a great influence on the development of the human style of management. The Hawthorne studies concluded that the informal work group had an influence on worker productivity. Maslow's hierarchy of needs was developed during this time span and is still used widely today. The human relations management style considers the informal work group and emphasizes employee welfare, motivation, and communications.

- *Bureaucracy*: Bureaucracy signifies the advent of a structured and formalized management network. Just as the workers have been given a rigid structure to follow, so have the management personnel of the organization. The loss of a manager does not cause a drastic loss to the organization; another individual could step in and perform the function with only minor adjustments, analogous to the replacement of an hourly employee, using this model. The larger corporations and governmental units are associated with this style. The decision-making powers are concentrated in the higher echelons of the organization. Most federal and state governments are monuments to the ideals of bureaucracy.

- *Quantitative Management*: The end of World War II resulted in the application of more mathematical analysis to the problems of industry. Formal approaches to solving problems were developed. The use of quantitative methods in problem solving was emphasized. The methods included statistical theory, linear programming, queuing theory, simulation, production control, and quality. The rise of management is an outshoot of this technique.

- *Organizational Behavior*: Organizational behavior emphasizes those management activities that promote employee effectiveness. McGregor (1960) formulated his Theory X and Theory Y during the 1950s and published them in 1960. However, they still exert a tremendous influence on many people. Recently, there has been more study on leadership, employee involvement, and productivity. This approach places emphasis on leadership skills, including situational leadership; for example, coaching, facilitating, etc.

- *Systems Theory*: The emphasis on looking at the organization as a whole emerged during the late 1950s. Instead of stressing one part of the organization and ignoring the effect on a secondary function, a view of total effect was considered. This was the start of the recognition of interaction with the outside environment. The quality movement seems to mirror this trend. Early efforts in Japan focused on the local area, followed by company-wide quality efforts, and now the emphasis is on supplier/customer relationships. Peter Senge's (1990) book, *The Fifth Discipline*, is a widely acclaimed systems book. The recent emphasis on total quality could fall into this category.

- *Contingency Theory*: Contingency theory advocates a flexible approach to the management of a firm. With the changing environment, at an ever increasing rate, there is not a single best way to manage the company. There may be key situational conditions that describe the industry or market that the company is in: low cost, high quality, high growth, etc. These would be the contingencies of a company. The managers need to analyze the situation and make decisions based on contingencies.

Reviewing the myriad methods of management, not all companies have advanced to the more current theories. Many management styles are based on theories developed 70 to 100 years ago. Current management theories often take 30 years to migrate throughout industry before they are commonplace. Some of the modern ways of managing are based on the theories of:

- *Douglas McGregor—Theory X and Theory Y*: Theory X and theory Y are the dominant theories in industrial practice readily recognized by most individuals. Douglas McGregor (1960), in his classic book *The Human Side of Enterprise*, pointed out the two sides of looking at how to manage employees. He referred to the sides as theory X and theory Y. The theories are based on how managers view their employees and how they would be treated for motivational and work-related purposes. Table 2.2 summarizes the two theories based on Schermerhorn (1993) and the Council of Indiana (2000).

  Some organizations may seem to fit one of the two modes, or it may be that certain managers reflect a different philosophy than that of the organization. Certainly it appears that the attitude or beliefs of the general manager will influence the workings within an organization. The theory X manager will be more control oriented, very directive, and narrow in the approach. The subordinates will need to be more passive, dependent, and reluctant in their ways.

TABLE 2.2

Summary of Theories X and Y

Theory X	Theory Y
Dislike work	Willing to work
Lack ambition	Accepts responsibility
Is irresponsible	Has self-direction
Won't change	Has self-control
A follower	Highly capable

This will reinforce the theory X manager's thoughts and expectations of the employees.

The theory Y manager will allow more participation, freedom, and responsibility. This will allow the subordinates to be more receptive and responsible to the nature of the work. This should increase the opportunities for activating the higher level needs of employees and produce better employees and results. Again, the self-fulfilling prophesy will be met.

- *William Ouchi—Theory Z*: Schermerhorn (1993) outlined the start of theory Z. William Ouchi (1981) published a book on the subject: *Theory Z: How American Business Can Meet the Japanese Challenge.* The book emphasized the Japanese approaches to management techniques. Some of the basic principles of theory Z follow:
  - Long-term employment philosophy
  - Slower promotions and more lateral job movements
  - Emphasis on career planning and development
  - Consensus decision making
  - Employee involvement
  - Continual emphasis on quality improvement
  - Emphasis on the role of the work group

Theory Z appears to be an adoption of theory Y concepts, put into practice at the organizational level. The whole-scale adoption of any management theory, Japanese, German, or American, should be based on the merits of the system involved.

Leadership, on the other hand, must be understood in three converging approaches: (1) qualities—what you are, (2) situational—what you know, and (3) functional—what you do.

Leaders have always played a primordial emotional role. That is, whether tribal chieftain or *shamaness* (a priestess in shamanism: the title is used interchangeably to designate certain individuals possessing magico-religious powers and found in all "primitive" societies), they earned their place in large part because their leadership style was emotionally compelling. We see throughout human history that great leaders have moved people to do things. They seem to ignite our passion and inspire the best in us. Yet, when we try to explain why they are so effective, we speak of strategy, vision, or powerful ideas. But the reality is much more primal: Great leadership works through our emotions. In fact, that is the reason that great leaders can create a positive emotional climate and enjoy sustained business performance in the organization that they are affiliated with. They do that by creating a resonance of positivity in everything that they do.

Etymologically, *resonance* is rooted in the Latin word *resonare*, which means "to resound." The *Oxford English Dictionary* (1989) defines resonance as "the reinforcement or prolongation of sound by reflection" or, more specifically, "by synchronous vibration." The human analog of synchronous vibration occurs when two people are on the same wavelength emotionally—when they feel in sync. And true to the original meaning of resonance, that synchrony resounds, prolonging the positive emotional pitch.

The phrase "too many chiefs and not enough Indians" needs no explanation, but in addition to its ethnic insensitivity, the statement misses from a managerial standpoint. In organizations aspiring to growth and continual improvement, relationships are more complex and options more numerous than the either/or dictate implied by the notion of leaders and followers.

Virtually no one leads all of the time. Leaders also function as followers; everyone spends a portion of his day following and another portion leading. For example, a senior vice president may fill the role of powerful person when dealing with subordinates but not when dealing with the company president. In turn, a company president who refuses to respond to others will soon earn the ire of his or her board or a legion of customers (Townsend & Gebhardt 1997).

More illuminating is to consider the relationship between followership and leadership as two points on a continuum, anchored on one end by "passive followership"—a phrase from Robert Kelley's book, *The Power of Followership*, and roughly equivalent to slug behavior—and on the other by "capital-L leadership"—a phrase from *Five-Star Leadership* and roughly equivalent to a prophet's charisma. In between, moving from passive followership to capital-L leadership, a person passes through "active followership" and small-l leadership (Townsend & Gebhardt 1999).

Most people spend their lives moving back and forth between these latter two. Active followers and small-l leaders contribute a great deal to those occasions when everyone "just pitches in and gets it done." That's known as *teamwork*. At these times, leader and follower roles change so frequently that they are not worth labeling. When properly prepared, people who do not normally consider themselves leaders will assume the role whenever their leadership skills are called for.

Unfortunately, most American businesses concentrate their training efforts (if they train for leadership) on capital-L leadership. It is, after all, the more exciting option, the sort of leadership exercised by women and men whose decision-making power has far-reaching impact. It is the stuff of which legends are made. Small-l leadership, on the other hand, involves decisions with immediate impact, usually on people known to the decision maker.

Neither form of leadership operates in a vacuum. Habitual mistakes of capital-L leadership easily obliterate the bottom-line impact of small-l leadership decisions; lackadaisical acts of small-l leadership can easily nullify capital-L leadership decisions. Any leadership curriculum that glosses over small-l leadership creates a void as do curricula that overlook followership. If a person has not been trained, formally or informally, to fill the follower role, the odds are significantly lower that he or she will ever reach his or her leadership potential (Townsend & Gebhardt 1999).

A look at an article written by Sgt. 1st Class Michael T. Woodward for the *U.S. Army's Infantry Magazine* in mid-1975 suggests the scope of a followership course. Woodward points out that followers need to commit to the organization's mission, which in turn requires that they understand the mission and concur with its aims. This simple idea is, of course, a major stumbling block in organizations that demand blind obedience from lower-level employees. In fact, creating an environment in which employees become active, committed followers requires real effort on all sides and more than a modicum of trust. Woodward's goal was to create competent followers able to estimate the proper action required to contribute to mission performance and, in the absence of orders, to take that action.

Woodward included 10 guidelines for followers in his article. He used the U.S. Army's leadership principles as a point of reference, and the list reflects how close active followership is to small-l leadership:

1. Know yourself and seek self-improvement.
2. Be technically and tactically proficient.

3. Comply with orders and initiate appropriate actions in the absence of orders.
4. Develop a sense of responsibility and take responsibility for your actions.
5. Make sound and timely decisions and recommendations.
6. Set the example for others.
7. Be familiar with your leader and his or her job and anticipate his or her requirements.
8. Keep your leaders informed.
9. Understand the task and ethically accomplish it.
10. Be a team member—but not a yes man.

"Effective leadership requires followers who are more than Pavlovian reactors to their leaders' influences," noted Woodward. "When followers actively contribute, are aware of their function and take personal pride in the art of followership, then the joint purpose of leadership and followership—higher levels of mission accomplishment—is achieved effectively. Professionalism in followership is as important in the military service as professionalism in leadership (p. 6)."

Educating people to help them become productive followers and leaders is an important leadership responsibility. Thus, any thoughtful leader has three top priorities:

- Accomplish the mission.
- Take care of the people in the organization.
- Create more leaders.

An organization's senior management must be willing not only to invest money and resources but take part in discussions and lead by example. They must come to grips with the continuum from followership to leadership, rather than present the two as opposing concepts across a yawning gap. Otherwise, folks caught on the chasm's followership side will show no ambition to become leaders. The distance will prove too great for all but the greatest leaps of faith.

Over the years, many articles and books have been written on leaders and leadership. One may summarize the findings of these writings in the following leadership behaviors:

- Business acumen—know-how that moves the organization/company forward
    - Industry knowledge—understands the industry and the factors that can affect regional and corporate goals; uses industry knowledge in planning and decision-making
    - Chooses appropriate action—formulates clear decision criteria; evaluates options by considering implications and consequences; chooses an effective option
    - Keeps up to date—stays abreast of current developments and trends in all relevant technical/professional knowledge areas
    - Involves others—includes others in the decision-making process as warranted to obtain good information; makes the most appropriate decisions and ensures buy-in and understanding of resulting decisions
    - Entrepreneurial insight—seizes opportunities to increase current business or to expand into new markets, products, or services
    - Generates alternatives—creates relevant options for addressing problems/opportunities and achieving desired outcomes
    - Commits to action—makes decisions within a reasonable time
- Develops employees and teams—fosters teamwork
    - Clarifies current situation—clarifies expected behaviors, knowledge, and level of proficiency by seeking and giving information and checking for understanding
    - Explains and demonstrates—provides instruction, positive models, and opportunities for observation in order to help others develop the skills; encourages questions to ensure understanding
    - Provides feedback and reinforcement—gives timely, appropriate feedback on performance; reinforces efforts and progress
    - Volunteers assistance—offers to help others achieve mutual goals
    - Subordinates personal goals—places higher priority on team or organization goals than on own goals
    - Involves others—listens to and fully involves others in team decisions and actions; values and uses individual differences and talents
    - Models commitment—adheres to the team's expectations and guidelines; fulfills team responsibilities; demonstrates personal commitment to the team

- Connects with customers—views the customer as a priority
  - Acknowledges the person—greets customers promptly and courteously; gives full attention
  - Clarifies the current situation—asks questions to determine needs; listens carefully; provides appropriate information; summarizes to check understanding
  - Meets or exceeds needs—acts promptly in routine situations; agrees on a clear course of action in nonroutine situations; takes opportunities to exceed expectations without making unreasonable commitments
  - Confirms satisfaction—asks questions to check for satisfaction; commits to follow-through, if appropriate; thanks the customer
  - Takes the "heat"—handles upset customers by hearing the customer out, empathizing, and apologizing
  - Maintains audience's attention—keeps the audience engaged through use of techniques such as analogies, illustrations, body language, and voice inflection
  - Adjusts to the audience—frames message in line with audience experience, background, and expectations; uses terms, examples, and analogies that are meaningful to the audience
  - Comprehends communication from others—attends to messages from others; correctly interprets messages and reasons appropriately
  - Develops own and others' ideas—presents own ideas; seeks and develops suggestions of others; makes procedural suggestions
  - Facilitates agreement—uses appropriate influence strategies to gain genuine agreement; persists by using different approaches as needed to gain agreement
  - Builds rapport—makes favorable impressions by interacting with prospects/clients in a manner that builds effective relationships
- Drive for results—sticks with it to get the job done
  - Persists in efforts—works to achieve goals in spite of barriers or difficulties; actively works to overcome obstacles by changing strategies, doubling efforts, using multiple approaches, etc.
  - Redirects focus—adjusts focus when it becomes obvious that a goal cannot be achieved; redirects energy into related achievable goals if appropriate
  - Prioritizes—identifies more critical and less critical activities and assignments; adjusts priorities when appropriate

- Determines tasks and resources—determines project/assignment requirements by breaking them down into tasks; identifying equipment, materials, and people needed and coordinating with internal and external partners
- Schedules—allocates appropriate amount of time for completing own and others' work; avoids scheduling conflicts; develops timelines and milestones
- Leverages resources—takes advantage of available resources (individuals, processes, departments, and tools) to complete work efficiently
- Stays focused—uses time effectively and prevents irrelevant issues or distractions from interfering with work completion

Because these behaviors have become the standard bearers of good leadership in any organization, we must be cognizant that sometimes obvious, often overlooked, the lessons of leadership must be managed and respected by those who help guide the way in business. With this in mind let us review history for some classical examples that might give us some insights:

- Rembrandt, an artist, has taught us to see the whole picture rather than focus too close.
- Julius Caesar, famous for his Roman justice during his many campaigns, taught us to be pragmatic and, of course, if we need to bluff our opponents we must make sure that it is a bluff before we call it. We must look not just to the clear path ahead but to what lies beyond. Nothing else will do. If the leader loses his perspective he may be ambushed by an enemy who was not expected. Therefore, as a leader you must assess the situation and use the appropriate and applicable resources to complete the task at hand. If not, your efforts will be counterproductive.
- The French knights at the Battle of Crecy taught us that courage alone is not enough. We must have appropriate and applicable knowledge and power to overcome the situation.
- *The Prince*, written by Machiavelli in the early sixteenth century, gives us another look at leadership. Machiavelli was right on the mark when he alluded to the fact that the truth may be unsavory but that is no reason to discount it. In modern days we can see this in the application of reengineering and process mapping in pursuit of

finding optimum processes. In both cases, the leader's power comes from those who actually do the work. Therefore, we can interpret Machiavelli's plea that raw leadership is a cry in the wilderness without the troops both following and contributing. After all, the core function of leadership is the advancement of the greater good. As the roots that help feed the tree, the people of the enterprise and the structure that makes their success possible should be high priorities for their leader.

- Of course, we must not forget the innovation by Wal-Mart's leadership to cross-dock the distribution centers and thereby get ahead of the discount K-Mart empire. For Wal-Mart's leadership, discounting was more than cheap prices. It turned out to be a venture and a vision than turned customer satisfaction into a phenomenal success and the envy of many.

- Finally, leadership should be questioned when things appear to be going well. This lesson was learned the hard way in the case of Hewlett Packard in 2005 when the board fired CEO Carly Fiorina. On the other hand, James Clavell's 1962 novel *King Rat*, set in a prisoner of war camp during World War II, articulated the autocratic leadership principle dramatically. He was able to demonstrate that an autocratic leader can appear very functional in a crisis. However, that leadership is rarely sustainable.

Many leaders in a variety of organizations (both profit and not-for-profit) confuse *efficiency* (the use [allocation] of internal resources) and *effectiveness* (issues that concern customer specifications; the fundamental question here is how the organization satisfies the customer's requirements). To satisfy these two issues, the leaders arm themselves with special analysts (human middleware) to relate or to explain things and their relationship to the overall performance of the organization.

For high-performance quality, the leader in any organization should invest and expect measurable items that should contribute *total improvement* to the entire organization at large. Table 2.3 displays some of the items that a leader should invest in and expect.

Unequivocally, leaders have a tremendous influence in any organization. However, their responsibility is much more than influence. It has to do with managing the whole organization using the systems approach as opposed to focusing on suboptimization, which deals with improvements

TABLE 2.3

Items That a Leader Should Invest and Expect

Invest	Expect
Talent management	More flexibility and improved capabilities. Always make it a policy to hire for talent. Move people around and into the business
Accounting systems	Improved control over data
Internal controls	Assure integrity of data and results
Performance scorecards	Prioritization of what matters most. Zero in on activities that create value for the business and the customer
Business intelligence	Better insights, smarter decisions with less risk
Merger integration	Day-one readiness
Technology innovations	Appropriate and applicable integration of new technologies that can build specific business needs
Tax and treasury	Higher profits and better compliance
Quality initiatives	Overall continual improvement and customer satisfaction

in specific areas of the organization at the expense of other departments within the organization.

In effect, leaders can become true architects for improving, changing, and redefining success in organizations if they use their vision, conviction, and the appropriate tools to implement their goals in such a way that will benefit the organization itself, as well as its employees, suppliers, and customers.

Leadership, therefore, is a bundle of characteristics that, once understood, can be improved. Toward this end the Myers–Briggs-type indicator (MBTI) characteristic assessment was developed by Katharine Briggs and her daughter, Isabel Briggs Myers (Myers and Myers 1987; Hirsh and Kumnerow 1990). Their assessment tool was based on the prior work of C. G. Jung, a Swiss psychiatrist, who had studied people's behaviors for years.

The MBTI assessment tool is primarily concerned with the valuable differences in people that result from where they like to focus their attention, the way they prefer information, the way they make decisions, and the kind of lifestyles they adopt. Each of the combination types has its own set of inherent strengths. Table 2.4 shows the two preference options for each of the four characteristic scales. There are 16 possible characteristic combinations, two of which are detailed in Table 2.4.

In summary, one may surmise that the two major combinations are the ISFP and the ENTJ. Each one has its own characteristics and obviously the other combinations offer their own. Here we present only the two.

TABLE 2.4

Myers-Briggs–Type Indicator Characteristic Scales

Where do you prefer to focus your attention? The EI scale	
Extroverts (E) tend to focus on and be energized by the outer world of people and things. Extraverts prefer to communicate more by talking than writing. They like action and variety.	Introverts (I) focus on their inner world. Introverts tend to be more interested and comfortable when they can work quietly without interruption. They need time to reflect before acting.

How do you acquire information? The SN scale	
Sensors (S) focus on the realities of a situation. They are realistic and practical. Sensors are good at working with a great number of facts. They are careful with detail.	Intuitives (N) go beyond the information obtained by their senses. They look at the big picture and overall patterns. They value imagination, inspiration, and possibilities.

How do you make decisions? The TF scale	
Thinking (T) types make decisions objectively, on the basis of cause and effect. They focus on the logical consequences of any choice or action. They are good at analyzing what is wrong with something.	Feeling (F) types make decisions based on internally centered values. They consider how important the choices are to themselves and others. They are sympathetic, appreciative, and tactful.

How do you orient toward the outer world? The JP scale	
Judges (J) like to live in a planned, orderly fashion, wanting to regulate life and control it. They want to make quick decisions and come to closure. They are structured and organized.	Perceivers (P) live in a spontaneous, flexible way, gathering information and keeping options open. They seek to understand life rather than control it. They trust their ability to adapt.

1. ISFP: Retiring, quietly friendly, sensitive, kind, modest about their abilities. Shun disagreements, do not force their opinions or values on others. Usually do not care to lead but are often loyal followers. Often relaxed about getting things done.

2. ENTJ: Frank, decisive leaders in activities. Usually good in anything that requires reasoning and intelligent talk, such as public speaking. Usually well informed and enjoy adding to their fund of knowledge. May appear more confident than their experience in an area warrants.

As the Council of Indiana (2000) has pointed out, it is imperative that we all understand that there are no good or bad individual types and there are no better or worse combinations of types in relationships. All of us use all of the preferences at different times. An individual type generally is made up of those preferences used most frequently. Type does not explain everything, and it does not measure abilities. Taken to its optimum, the MBTI can be used by individuals in making career choices and in understanding themselves and their behaviors. Organizations use the MBTI by making the most of human resources, improving teamwork, resolving conflict, and adapting management styles.

The importance of understanding the different styles is that leaders at all levels of the organization are expected to interact with senior managers, middle managers, or hourly employees. Each leader will develop his or her own approach to managing people, within the confines of the organization. Here we must emphasize that quite often the organization itself may reflect one of the previously discussed management styles in which the leader must operate.

In the managing of people on a day-to-day basis, leaders and/or managers have the ability to choose from among many styles. These choices range from the autocratic to democratic. Autocratic leaders make decisions on their own and then announce them to the work group. Democratic leaders solicit input from their group, have joint discussions with the group, and attempt to arrive at a consensus decision. The decision reached by an autocrat leader is fast but it may take a long time to be implemented. If it is implemented in a short time, there are individuals who will not agree with it. On the other hand, a democratic decision may take longer to be reached but it will need a much shorter implementation cycle because everyone will have a chance to participate in the decision-making process. Individual leaders may use different management styles; some of the most common ones are summarized here based on Council of Indiana (2000):

- *Manage by facts*: Information must be gathered from the applicable work situation and analyzed to generate recommendations. This concept is from the Japanese quality teachings and in essence it is a policy of making decisions on gathered data. Thus, if a problem exists on the work floor, use any or all of the methodologies and/ or tools appropriate and applicable to the situation to gather data. Geneen (1984) made mention of all decisions being made only after the "facts" were gathered. Ishikawa (1982) also discussed the use of data for decision making. The facts-based management style is to manage through collection of data and to remove the distortion of everyday noise from the decision-making process.

- *Situational leadership*: For effective managing or leadership, the situation is analyzed first, before deciding what to do. The style of a leader can change, depending upon the situation. For example, in an emergency situation, the proper style may be an autocratic one, in which the leader makes all the decisions; in a very calm and peaceful moment with nothing at stake, the leader can let chaos rule; in a teaching or training environment, it may be best to let the trainees guide themselves, with the leader providing guidance or playing a facilitator role.

  Hersey and Blanchard (1982) developed a very popular situational leadership model. In that model, the leader will change roles depending on the subordinate's skill level. In the Hersey and Blanchard situational leadership model, the leader will use one or more of the following situational leadership styles, depending upon the circumstances. That is, (a) directive—telling and directing what to do; (b) selling—coaching, helping the employee learn, yet still directing their actions; (c) participating—the employee is skilled in his work but still needs guidance; and (d) delegating—the employee is now skilled and able to do the work.

  Harrington (1995), on the other hand, described five types of management styles, which have some similarity to Herseys: (a) Coach—directing and showing the employee how to do the job, and includes an emphasis on minimizing errors and cheerleading the employee to a successful task; (b) Teacher—teaching the employees concepts, measuring performance, revealing errors, and helping them to succeed; (c) Boss—providing assignments, seeking completion of assignments, developing and training employees for quality and productivity; (d) Leader—delegating accountability and responsibility for work to the employees, and acting as a coordinator of employee efforts, removing obstacles in their way; and (e) Friend—delegating work to employees and maintaining personal relationships with them, while working together to solve problems and allowing freedom of decision making.

- *Empowerment*: Empowerment is the process of sharing power with the employees. If empowered, the employees will take more initiative in their jobs and in the company. They will share in the decision making. In a team environment, the leader can act as the coach for the team, providing guidance on key issues.

Both leaders and managers are guided in their day-to-day operating modes from their initial behavioral background. Social scientists such as Elton Mayo (1949) (Hawthorne studies), Abraham Maslow (1943), Douglas McGregor (1960), David McClelland (1988), Frederick Herzberg (1993, 2008), Clayton Alderfer (1969), and others have attempted to sort out the human factors of management to produce a better workplace. The leader's or manager's thinking and style of managing are usually characterized by one of these theories, which are based on behavioral science. However, fundamental to all is what Schermerhorn identified as the principal assumption that states: "People are social and self-actualizing. The branches of the behavioral approach to leadership or management share a belief in the social and self-actualizing nature of people. People at work are assumed to act on the basis of desires for satisfying social relationships, responsiveness to group pressures, and the search for personal fulfillment (1993, p. 124)."

McClelland (1962) went a step further and proposed a set of motivators that individuals acquire from social interaction or culture. He called them *learned needs* and they are as follows:

- Need for achievement
- Need for affiliation
- Need for power

McClelland (1962) proposed that an individual will strive to fulfill the most strongly based need. For example, the high achievement individual will work hard on challenging goals and opportunities. The person high in need for power will enjoy situations of authority and responsibility. The person with high affiliation need will have a need for acceptance by a group. On the other hand, Ivancevich and Matteson (1996) claimed that situations requiring strong leadership and management decisions that negatively affect the group might be uncomfortable for an individual with high affiliation needs.

The lessons of history and the examples of our present-day peers, it is hoped, offer ample guidance. A true leader will embrace it.

Finally, Peter Drucker, the late renowned management consultant, made the observation that the leader of the past may have been a person who knew how to tell, but the leader of the future will be a person who knows how to ask (2004). Too few leaders lead with questions. They tend to dictate or debate rather than inquire and discuss.

This is an unfortunate situation, because with the growing complexity and speed of change in the world today, a leader simply will not know enough to adequately tell his subordinates or colleagues what to do. In today's turbocharged environment, no one can master all of the data needed to address all globally complex problems. Asking the right questions depends on what you want to accomplish. To paraphrase Marquardt (2007), the chosen questions must inspire people to act in new ways, expand their range of vision, and enable them to contribute more to the organization.

Powerful questions are usually open-ended—that means that they are not biased or leading questions and certainly they are not looking for a specific answer. They often begin with why, how, or what do you think about … ? They help people to discover answers, which develops responsibility and reestablishes authority and ownership. A good leader or manager should avoid the use of confrontational questions—ones that are negative, put people into a defensive mode, and drain energy. Asking "What's wrong?" threatens self-esteem and discourages honesty, creativity, and collaboration.

So instead of asking what went wrong, ask questions that focus on what has gone well, what could be done, and how it could be improved. The focus remains on improvement and continuous learning rather than complaining and venting. By being open-minded and positive, you encourage a broader range of responses.

An excellent leader must always remember that great questions empower people and instill in them a sense of their own strength and efficacy. When you are truly asking, you are sending the message that the subordinate's ideas are good or maybe even better than your own. In addition to conveying respect, it encourages the person's development as a thinker and problem solver.

Effective types of questions are as follows:

- Create clarity (Can you explain what happened?)
- Construct better working and personal relations (How have sales been going?)
- Help people think analytically and critically (What are the consequences of going this route?)
- Inspire people to reflect and see things in fresh, unpredictable ways (Why does this strategy always work?)
- Encourage breakthrough thinking (Can that be improved?)
- Challenge assumptions (Why do we always choose this method?)

- Create ownership of solutions. (Based upon your experience, what do you suggest we do here?)

On the other hand, there are questions that leaders should avoid. For example, questions that focus on why a person did not and cannot succeed force him or her to take a defensive or reactive stance and strip him or her of power. Such questions shut down opportunities for success and do not allow people to clarify misunderstandings or to achieve goals. Other typical questions include the following:

- Why are you late in delivering this product?
- What is your problem in implementing this program within budget?
- Who is not doing their part?
- Don't you know any better than that?

Likewise, leading questions that seek a specific answer put the other person in a negative light, push the questioner's agenda, and exert social pressure to force agreement. These types of questions—You wanted to do this by yourself, didn't you? Don't you agree that John is the problem? or Everyone else thinks this is the proper analysis of the situation, why don't you?—inhibit people from answering candidly and stifle honest discussion.

Leaders who are unaware of the potential of questions needlessly engage in a fractious, pressure-filled existence. Leaders who lead with questions will create a more humane workplace as well as a more successful business. Leaders who use questions successfully will truly empower people and change organizations.

SUMMARY

In this chapter we have addressed leadership as it affects quality in any organization. We focused on key characteristics for understanding leadership on the one hand and management on the other. In the next chapter we will address the execution phase. That means how a leaders proceed to execute a decision and infuse it into the entire organization.

REFERENCES

Alderfer, C. (1969). "An empirical test of a new theory of human needs." *Organizational Behavior & Human Performance,* 4(2), 142–175.

Clavell, J. (1962). *King Rat.* NY: Bentam Dell.

Council of Indiana. (2000). *CQM primer,* 4th ed. West Terre Haute, IN: Council of Indiana.

Delavigne, K. T. and J. D. Robertson. (1994). *Deming's profound changes.* Englewood Cliffs, NJ: Prentice Hall.

Drucker, P. (2004). http://www.forbes.com/2004/11/19/cz_rk_1119drucker.html

Galbraith, J. (1995). *Designing organizations: An executive briefing on strategy, structure, and process.* San Francisco: Jossey-Bass.

Geneen, H. S. (1984). *Managing.* Garden City, NY: Doubleday.

Harrington, H. J. (1995). *Total improvement management.* New York: McGraw-Hill.

Hawthorne studies. http://www.analytictech.com/mb021/handouts/bank_wiring.htm

Hersey, P. and K. Blanchard. (1982). *Management of organizational behavior utilizing human resources,* 4th ed. Englewood Cliffs, NJ: Prentice Hall.

Herzberg,F. (1993). *Motivation to Work.* Piscataway, New Jersey: Transaction Publishers.

Herzberg,F.(2008). *One More Time: How Do You Motivate Employees?* New York: Harvard Business Press.

Hirsh, S. K. and J. M. Kumnerow. (1990). *Introduction to type in organizations.* Palo Alto, CA: Consulting Psychologist Press.

Ishikawa, K. (1982). *Guide to quality control.* White Plains, NY: Quality Resources.

Ivancevich, J. and M. Matteson. (1996). *Organizational behavior and management,* 4th ed. Chicago: Irwin.

Kelley, R. (1992). *The Power of Followership.* New York: Doubleday Business.

Kotter, J. P. (1982). "What effective general managers really do." *Harvard Business Review* December, 21–32.

Kotter, J. P. (1990). "What leaders really do." *Harvard Business Review* 2(3): 12–16.

Machiavelli, N. (1984). *The Prince.* New York: Bantam Classics.

Marquardt, M. J. (2007). "The power of great questions." *T&D* February, 92–93.

Maslow,A. H. (1943). A Theory of Human Motivation. *Psychological Review,* 50(4), 370–396.

Mayo, E. (Hawthorne studies). (1949). *Hawthorne and the Western Electric Company: The Social Problems of an Industrial Civilization.* New York: Routledge.

McClelland, D. (1962). "Business drive and national achievement." *Harvard Business Review* July–August, 24–28.

McClelland, D. (1988). *Human Motivation.* New York: Cambridge University Press.

McGregor, D. (1960). *The human side of enterprise.* New York: McGraw-Hill.

McGregor, D. (1960). *The Human Side to Enterprise.* New York: McGraw Hill Higher Education.

Myers, P. B. and K. D. Myers. (1987). *Myers-Briggs type indicator.* Palo Alto, CA: Consulting Psychologist Press.

Ouichi, W. (1981). *Theory Z: How American business can meet the Japanese challenge.* Reading, MA: Addison-Wesley.

The Oxford English Dictionary. (1989). 2nd ed. New York: Oxford University Press.

Schermerhorn, J. R., Jr. (1993). *Management for productivity,* 4th ed. New York: John Wiley & Sons.

Senge, P. M. (1990). *The fifth discipline.* New York: Doubleday.

Strauss, L. (1970). *Xenophon's Socratic Discourse: An Interpretation of the "Oeconomicus."* Ithaca, NY: Cornell University Press.

Taylor, F. W. (1911). *The principles of scientific management.* New York: Harper and Brothers.

Townsend, P. and J. Gebhardt. (December 1997). "Followership: An Essential Element of Leadership." *Quality Digest*, p. 11. http://www.qualitydigest.com/dec97/html/townsnd.html

Townsend, P. and J. Gebhardt. (1999). *Five Star of Leadership: The Art and Strategy of Creating Leaders at Every Level.* New York: J. Wiley and Sons.

Woodward, M. T. (July-August 1975). *U.S. Army's Infantry Magazine.* pp. 4–6.

SELECTED BIBLIOGRAPHY

Adair, J. (2003). *The inspirational leader.* London: Kegan Page.

Cohen, E. (2008). "Welcome to the new global frontier." *T&D* February, 50–55.

Faigenbaum, A. V. (2007). "The international growth of quality." *Quality Progress* February, 36–40.

Heil, G., T. Parker, and R. Tate. (1994). *Leadership and the customer revolution.* New York: Van Nostrand Reinhold.

Hollander, E. P. (1992). "The essential interdependence of leadership and followership." *American Psychological Society* 1(2): 71–74.

Malone, R. (2006). "Studies in leadership." *Manufacturing Automation* March, 66.

Merriam-Webster Online Dictionary. (2005). "Follower." Available at: http://www.m-w.com/dictionary/follower (accessed 11 December 2005).

Pace, A. (2008). "The slippery slope of leadership." *T&D* October, 10–11.

Phillips-Donaldson, D. (2002). "On leadership." *Quality Progress* August, 24–25.

Stamatis, D. H. (1997). *TQM engineering handbook.* New York: Marcel Dekker.

White, H. S. (1987). "Oh, where have all the leaders gone?" *Library Journal* 112(16): 68–69.

3

Execution

In the last chapter we discussed the characteristics of a leader and how those characteristics affect the organization. In this chapter we focus on execution of decisions. What is it about the leader that makes decisions flow through the organization with good results? Why is it important to have an execution policy based on schedule, budget, and appropriate training for all concerned? These questions and more are addressed from a strategic planning perspective.

A recent global study conducted by the HR Institute and the American Management Association found that execution derailments are due to the following (2006):

- Heavy reliance upon annual strategic and performance review
- Slow and ineffective decision making
- Weak employee engagement
- Insufficient attention to customer needs
- Murky roles, responsibilities, accountabilities, and progress measures
- Inability to work laterally across functions and departments
- Weak monitoring of progress
- Ineffective delegation of execution to others

These findings point to the need for a new level of coordination, improved integration between external and internal realities, more frequent adjustments, and a more systematic determination of how to align the organization to its strategy.

The disconnect between strategy and implementation can be bridged by inserting a new planning step between strategic planning and project management (PM). This new step, called *execution planning*, is a creative process to identify, communicate, and implement initiatives to achieve strategic goals.

Most organizations, in the rush to act on strategy, pay too little attention to finding the best implementation initiatives. Shortcuts are taken, such as repackaging existing projects that appear to support a new strategy—which merely replicates what other companies have done without any customization. There is a clear parallel between our experience of using analysis and planning to improve quality and how execution planning can boost implementation success rates.

Executives can copy strategic plans, but they cannot duplicate execution. Consider the automotive industry. All manufacturers pursue the same strategy, yet few execute it well. Toyota opens its operations for observation, but few can replicate the company's results because it requires more than tools, techniques, and schedules. Execution planning addresses the intangibles of cross-functional integration, reward systems, and culture as well as the tangibles captured in most planning documents. Execution, not strategy, offers an exclusive competitive advantage.

Consider the costly history associated with implementing most enterprise resource planning systems or integrating human resource systems or other comprehensive information technology (IT) solutions. For many firms, false starts, delays, and confusion characterize implementation. The new planning step targets exactly these issues. Further, this new step ends the old ineffective practice of executives tossing a strategy "over the transom" for others to carry out. The bottom line is that this new step integrates implementation cross-functionally and across departments, increasing the chance for success. It is a sound upfront investment that prevents the confusion, rework, and delays that characterize many implementation efforts.

Another reason that strategic plans fail to produce results is that most organizations operate on an annual cycle for strategic review, performance review, and budget allocation. Execution planning provides an immediacy that is currently lacking in many organizations. When questions, problems, or issues arise, it also serves as an efficient way to resolve these issues with timely adjustments or clarifications.

NASA reported that rockets are off course more than 80 percent of the time (2008). They would never meet their intended destination without making the necessary adjustments. The same thing occurs during execution. Execution plans can and do go astray, but they can still be successful, as long as the variance is noted in time and adjustments are made to get back on course.

In summary, the new execution plan:

- Improves cross-functional integration during monitoring.
- Focuses on the intangible people and cultural issues that are known to derail both strategy and change.
- Provides a timely review and resolution path.
- Launches customized initiatives that gain strong internal support.
- Aligns work to meet goals and cuts wasteful activities.

A NEW MIND-SET

In 1941, although the admirals at Pearl Harbor were highly experienced, they failed to see the possibility of a surprise attack. A common worldview, or groupthink, contributed to their shared blind spot. Research conducted at Ohio State University, titled "Why Decisions Fail," shows that this phenomenon also permeates business. The study found that 80 percent of management decisions are made without ever considering an alternative (Nutt 2002, Hubbard 2009).

This decision-making mode might have been safe in the 1800s and early 1900s, when the pace of change was slower, but in our current dynamic business environment, it is dangerous. As Darwin's theory suggests, "It is not the strongest of the species that survive, or the most intelligent, but the one most responsive to change." A single viewpoint limits our ability to see trends and effectively adapt in a timely manner. To neutralize groupthink, multiple mind-sets are infused into execution planning so that new options can be explored. This broad framework provides insight through key execution priorities. The framework functions as a checklist to ensure that all contingencies are considered and that the best alternative is executed. It also helps to communicate the plan to others. When a plan reflects all priorities, a high level of acceptance is more likely.

Execution planning may not yet have the panache of strategy formulation, but it substantially increases the likelihood of solid and sustainable results. It is the steak to strategic planning's sizzle. It also offers human resources (HR) and learning professionals the opportunity to serve as execution planning facilitators and become valued business partners.

Execution planning and HR and learning professionals are a natural fit. HR and learning professionals understand their entire organization, develop leaders and skills, and are process masters. Accounting and sales serve as prime examples for how an expansion of roles into finance and marketing, respectively, led to an increase in stature as key business players. If HR and learning professionals reposition themselves as contributors to execution, they will become indispensable partners to the executive team.

Can HR and learning professionals perform this role? Yes. Their skill sets match those required for an execution planning facilitator, including the following:

- Mastery of systems and integration mechanisms
- Ability to assess and develop new competencies
- Knowledge of culture and change management
- Organizational analysis skills
- Coaching and facilitating skills
- Ability to forecast and prescribe
- Consulting and customer service skills
- Questioning skills

The following example demonstrates that this new role is not a matter of technical skill; rather, it requires *mind-set flexibility, effective questioning*, and an *awareness of process*. A large manufacturing firm had just adopted a growth strategy when one of its older plants was cited by the Environmental Protection Agency (EPA) for violating new pollution standards. Top management's initial thought was to spin the problem through public relations (PR), which would assert that the older plant should be grandfathered or given a period of time to comply. By relegating the issue to PR, top management could avoid any further distraction.

This was not necessarily the best tactic. With few questions before management started to investigate other options, the problem shifted from its original direction. Among the questions that were asked, a single question like, "Are there any potential connections between growth demands and the current regulatory citation?" was able to generate a list, including shutting the plant down, challenging the new requirements in court, retrofitting the plant to meet the new requirement, and replacing the plant's technology with new technology that would meet EPA standards. To encourage the exchange of ideas, the team was also asked to identify every potential response.

With options on the table, the discussion now turned to each of them. Everyone agreed that any action that would advance the growth strategy would be compelling. Pressed to list additional factors, the executive team identified legal costs, community and customer goodwill, employee retention, union acceptance, meeting projected growth demands, initial cost outlay, tax savings, consumption of management time, competitive position, and support from the industry trade association, to name a few.

After careful evaluation, it became clear that the best decision was to introduce new plant technology that would also expand capacity to enable growth. What started out as a defensive public posture shifted in context and impact. The company was viewed as a responsible corporate citizen while it increased capacity, reduced operating costs, and prepared for growth.

The kickoff was coolly received by employees because the new technology would affect the size and skill of the workforce. The union resisted a reduction in force, but an agreement was easily negotiated when a no-lay-off policy was offered and a training plan crafted. During a temporary closure of the old facility, a new quality team was established and put through training to prepare for a vigorous process improvement effort when the plant reopened. And it did. Costs per unit fell 30 percent. The entire plant transformation was documented, and best practices were identified with the expectation that other plants would adopt the new technology.

This is one example of how HR and learning professionals can help organizations close the execution performance gap between strategic planning and implementation. What it takes is an understanding of what execution planning is, a framework to ensure comprehensive analysis, and an ability to ask questions using each priority in the framework. Table 3.1 shows some interesting rankings of employee attributes and mind-set requirements as well as priorities of recent college graduates and employers (Gaedeke & Tootelian 1989; Freestone, Thompson, & Williams 2006).

What can you do to start? A few suggestions:

- Look for an execution planning effort within HR to practice the new execution planning step.
- Conduct a "lessons learned" exercise on a recent execution effort to demonstrate how your knowledge can leverage better implementation.
- Partner with a colleague and prepare a model to use during execution planning.
- Reach out and offer assistance to guide a new execution initiative.

TABLE 3.1

Framework of Priorities

Employee Attribute	Student Rank	Employer Rank	Mindset Change Requirements	
			From	**To**
Oral communication skills	1	4	Event driven	Journey driven
Enthusiasm/motivation	2	1	Individual responsibility	Shared responsibility
Self-confidence	3	8	Informal process	Disciplined process management
Ambition	4	6	Person dependent	System dependent
Entrepreneurship	4	7	Philosophical beliefs	Rigorous applied practices
Initiative	6	3	Problem-solving environment	Process focused
Interpersonal skills	7	2	"Firefighting"	Handle problems upfront
Ability to articulate goals	8	9	Individual	Common understanding
Assertiveness	9	11	Understanding	Systematized knowledge
Written communication skills	10	14	Anecdotal knowledge	"Let's use it"
Maturity	11	5	"Just another initiative"	Leading indicators
Problem-solving skills	12	9	Traditional measures	Revenue enhancement focus
Leadership skills	13	12	Expense control focus	Tactical headcount addition
Related work experience	14	15	Headcount reduction	Work systems
Personal appearance	15	16	Individual applications	
Major area of concentration	16	13		
Creativity	17	17		
Quantitative skills	18	18		
Acknowledging limitations	19	19		

Execution offers HR and training the spotlight they need to prove value as business partners. All that is needed is the willingness to start executing.

PROJECT MANAGEMENT

When strategy fails to achieve expected results, is it because the strategy was ill conceived or the execution was flawed? Research is pointing more and more to problems with execution. It certainly is not a slam dunk. The failure to execute is a major concern of executives because it limits organizational growth, adaptability, and competitiveness.

Executives are not judged by the brilliance of their strategy but by their ability to implement it. The problem is how to close the gap between strategy and actual results (Lippitt 2007).

The solution is a new execution planning step. Not only will it solve the strategy execution problem, but it offers HR professionals a unique opportunity to assume a new organizational function and secure a seat at the decision-making table. That unique opportunity is project management.

Projects bring together resources, skills, technology, and ideas to achieve business objectives and deliver business benefits. Good PM helps to ensure that risks are identified and managed appropriately, and objectives and benefits are achieved within budget, within time and to the required quality.

The traditional knowledge areas of PM have always been based on time, cost, resources, and quality for ultimately a predefined customer satisfaction. Specifically, these knowledge areas are as follows:

- Project scope management
- Project integration management
- Project time management
- Project quality management
- Project human resource management
- Project procurement management
- Project risk management

Given this definition and parameters, have you ever thought about a strategy of success? The sad part of that is that 90 percent of all business owners and managers of large corporations do not plan. They react to failures (Lippitt 2007)! Yet, we all know that success is a matter of planning.

Planning is vital, but if you plan wrong, it could mean failure as well. So, what are we to do? On a personal level we must force ourselves to follow a time management regimen and in the large project domain we must follow the principles of PM.

The more we progress in technology, the more we find ourselves in need of shifting schedules and sorting our priorities. These new changes are indeed reorganizing our professional lives from the simple commonsense to very detailed, micromanaged items. Regardless of magnitude however, they all have something in common. That commonality is the global need of liberating us from anxiety caused by overwhelming deadlines.

The problem is that in all walks of life, including projects, no matter how well you compensate for your strengths and weaknesses, special circumstances will always occur. Can anyone control them? Well, not 100 percent. We can, however, plan accordingly and take our chances for success. Some of the simple things that we can do, on the individual level, are the following:

- *Stay ahead of the problem.* There is nothing better than appropriate and applicable preparation.
- *Manage your interruptions.* Time is always limited for everyone. Therefore, it is of paramount importance to organize your activities in blocks of time. For example, in the morning and late afternoon, take care of the correspondence; schedule meetings sometime before lunch; and so on.
- *Get out.* This is called *managed disruption.* Sometime within your daily schedule take a walk. It will give you an opportunity to rethink things over.
- *Design your own to-do list.* This is perhaps one of the best items to do in controlling time. Focus on no more than 10 items. Keep it simple and personalize it with specific questions about duration, interruptions, and level of priority.
- *Cat nap.* It is called the power nap. A light nap is refreshing and it gives you the opportunity to relax. Science suggests that making it a regular feature in your life will enhance your performance.

In our personal lives we can do something to minimize anxiety and meet our deadlines; there is also a discipline that helps us with large projects in dealing with quality, cost, and scheduling. That discipline is known as *project management.*

Project failures are all too common—some make the headlines, but the vast majorities are quickly forgotten. The reasons for failure are wide and varied. Some common causes are as follows:

- Lack of coordination of resources and activities.
- Lack of communication with interested parties, leading to products being delivered that are not what the customer wanted.
- Poor estimation of duration and costs, leading to projects taking more time and costing more money than expected.
- Insufficient measurables.
- Inadequate planning of resources, activities, and scheduling.
- Lack of control over progress so that projects do not reveal their exact status until too late.
- Lack of quality control, resulting in the delivery of products that are unacceptable or unusable.

PM is the approach that will eliminate or at least minimize the problems of cost overruns, missed deliveries, and poor quality issues. Without a PM method, those who commission a project, those who manage it, and those who work on it will have different ideas about how things should be organized and when the different aspects of the project will be completed. Those involved will not be clear about how much responsibility, authority, and accountability they have and, as a result, there will often be confusion surrounding the project. It must be emphasized here that PM is not an engineering discipline. Rather, it is methodology of controlling and measuring the allocation of resources (all resources) as they relate to a specific project.

Without a PM method, projects are rarely completed on time and within acceptable cost—this is especially true of large projects. A good PM method will guide the project through a controlled, well-managed, visible set of activities to achieve the desired results.

Obviously, not all projects are complicated and complex. On the other hand, all projects need some planning. Simple projects are common and they do not need extensive resources. In fact, they do not need specific guidelines of selection and priority policies. They are common events that happen daily in all sorts of organizations.

When projects are complex, large, and include other functional areas of the organization, PM is essential and some sort of regimen must be in place. So, what is this regimen process? At the very least it should include project charter, scope, deliverables, work steps, duration, and budget.

MANAGING QUALITY AND ITS RELATIONSHIP TO PROJECT MANAGEMENT

As everyone knows, *quality* is the degree to which customer needs are met or exceeded. *Quality management*, on the other hand, is a set of actions performed to bring about quality. With these two definitions, one can say that quality is as much a frame of reference and an attitude as it is a goal. Successful organizations treat quality management as an integral part of their operation, not as an afterthought. They methodically assess the causes of successes and failures and use the accumulated wisdom to improve and institutionalize the best practices (see Appendix D on the companion CD-ROM for a typical review for selected analyses).

Using the methodology of PM, organizations improve quality by clearly stating the project objectives during the early stages, setting integral review points within the project work plan, using convergence and iteration as a strategy to reach the quality needed, and providing the means to improve the use of methodology and tools in future projects.

For the purposes of the PM methodology, quality must be viewed from at least four perspectives:

1. *Measured from the process customer perspective.* One of the key components of a quality product is customer satisfaction. A flaw-lessly designed, defect-free product that does not meet the customer's needs cannot be considered to be high quality. To this end, the customer must be involved in both the quality management and development process.
2. *Viewed as both a process and a product.* A consistently high-quality product cannot be produced by a faulty process.
3. *Everyone's responsibility.* There must be a consistent and generalized set of principles that all parties can agree to and universally apply to make quality improvement a reality rather than simply a slogan.
4. *Measurable.* Without baselines, it is impossible to take advantage of any measurements. Without data, it is impossible to know how the progress of the project relates to those baselines.

By applying the principles of PM, one can see the key components for this methodology in the discipline of quality for the twenty-first century:

- *Early involvement in the process.* Early involvement by a quality advisor helps ensure the ability to meet customer needs and service expectations.
- *Help in selecting the right process.* There are many decisions to make when deciding what process to use to create a particular deliverable. The methodology is structured to allow users to select, based on specific objectives, the process they will use.
- *Guidance for empowering the team.* The methodology provides techniques, training, and clear-cut responsibilities for both the project team and the project manager.
- *A standardized way to perform the process.* Each route map provides a standard way to do the work, standard output, and a standard skill set necessary to achieve the process. This allows the work to be done the same way every time it is done.
- *A plan for integrating quality into the development process.* The project charter and the project plan both contain quality management components. In addition, quality checkpoints are built into the work plan to ensure a quality process and a quality product.
- *Continual involvement throughout the process.* Conducting periodic quality advisor reviews monitors performance of the project or program objectively and helps ensure consistency of deliverables.
- *A way to feed information back into the process.* At the end of each project, a feedback session allows the organization to report information from the project and enhance the enterprise knowledge base with best practice information. This enables continuous improvement of the PM process.

The methods for managing quality in a program are very similar to the ones used on a project. The key is to apply the methods effectively across the projects within a program. Quality reviews need to address the program process and products as well as the processes and products within individual projects. The program quality management plan should delineate the responsibilities for quality reviews within and across projects and provide the general standards for quality management.

If the project is local, one may also take into consideration whether the problems addressed are recurring; whether or not there is a narrow scope for improvement (either operational or financial); whether there are available and good data; whether or not the process is in control; whether or not the change will affect customer satisfaction; and whether or not the stakeholders have the power to make changes.

Though the approach is very powerful and indeed can provide positive results, there are situations in which PM may be used to confirm rather than develop a new approach. For example, if an approach exists, then one needs to review it and update it as necessary. If the project is part of a program, the program manager reviews the approach defined and applies it to the project as directed. If there has been a considerable span of time between the development of the initial project charter and startup of the current project, look for changes in the sponsor, business direction, project direction, or organization. Also consider changes and advances in technology that may affect the procedures and tools used in the project. If a standard approach exists for the enterprise, review it and apply that standard as required. (It is imperative to remember to note refinements to the enterprise-wide standards and approaches that may arise while applying them to the project or program.)

If the PM methodology is used to develop a quality management approach, then one must be aware to verify goals and objectives by ensuring the following:

- That you are in agreement with the sponsor of the project or program about the goals and objectives to be met by the program or project. Quality is measured through customer satisfaction; if you provide the right solution to the wrong problem, high quality will not have been achieved.
- That you create an overview describing the methodology quality management components, including quality reviews, standardization of output, standardization of tools, standardization of processes, standardization of skills, and integration of quality management principles in the development life cycle.

If the PM methodology is used to develop a quality management plan, one must be aware of the completeness and corrective criteria by making sure to:

- Establish completeness and correctness criteria for determining when each deliverable and its products meet technical specifications.
- Use these criteria as standards when conducting product reviews. The more succinctly you can define the criteria, the more likely you are to have successful deliverables that pass the quality review.

A third option of utilizing PM methodology is to use its principles and guidelines in making sure that the deliverable quality is in fact meeting customer requirements through reviews. The intent of the reviews is to ensure the consistency of the work within each deliverable and across deliverables. There are three major types of reviews:

1. *Team Review.* A brief meeting during which two or more team members examine the output of one or more tasks, focusing on correctness and completeness. A team review allows team members working on different aspects of a deliverable to confer and maintain consistency. If an inconsistency is discovered, the review offers a forum in which each member can contribute to the resolution.
2. *Formal Review.* A formal session to review a work product or deliverable, typically occurring at a major completion point (e.g., at the end of an activity or at the completion of a work product). This internal review typically involves participants who are directly involved in the development process. This review allows the work products to be reviewed in conjunction with other deliverables. Any defects found should be documented and shared with other members of the team. This may prevent the same defect from appearing in similar work.
3. *Management Review and Approval.* This external review point occurs at the end of each stage to verify that deliverables are complete, correct, and consistent before work proceeds. In addition to team members, this type of review involves participants who are external to the development process. The management review is conducted by members of the management and control structure and members of user management who are designated in the charter as approval authorities. When conducting the review, ensure that the person inspecting the deliverable knows what is expected of the review. If there are multiple inspectors, assign each inspector a different aspect of the deliverable. Errors found in a management review are defects; allow time for these defects to be corrected.

CLASSICAL PROJECT MANAGEMENT PROCESS

The process for executing any project to a successful conclusion in the classical approach is shown in Figure 3.1, which is made up of the following:

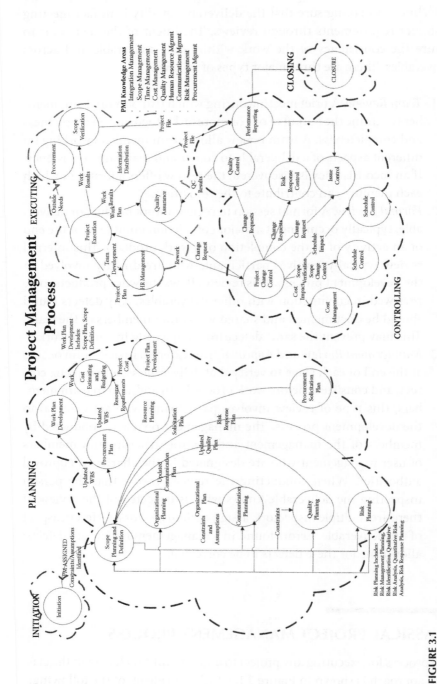

FIGURE 3.1

Classical project management process.

- *Initiation*: This is the area where the constraints and or assumptions are identified.
- *Planning*: This is the area where the work plan is developed.
- *Executing*: This is the area of carrying out the project as planned.
- *Controlling*: This is the area of making sure that planning and execution are in congruence. If not, appropriate adjustments must be made.
- *Closing*: This is the area where the project is completed and the transfer to the owners occurs.

RISK MANAGEMENT ACTIVITIES AS APPLIED TO PROJECT MANAGEMENT

Risk is a normal part of business and professional life. Nothing in the business world is straightforward or totally predictable. Change and external factors are always present to make us think better about the decisions we make. Risk needs understanding and managing for at least three reasons.

The first is an obvious one. By definition, professional life and business need to be somewhat conservative and defensive. After all, both are interested in protecting the current position in the marketplace and at the same time maximizing future opportunities.

The second is the need to embrace change and put in place strategic objectives for the development of the specific undertaking.

Thirdly, positively addressing risks can identify opportunities. Businesses and individuals who understand their risks better than their competitors can create market advantage.

The challenge of these three reasons, however, has always been for owner managers to find a comprehensive and robust process with which to manage and monitor the influencing factors in their business.

A coherent risk management plan plays a major role in providing an answer. It can help in both personal and business endeavors to avoid overextending ourselves, while at the same time coordinating opportunities and identifying risks on an informed basis. The plan must be backed fully by personal commitment (if personal) and by all partners and directors, as appropriate and applicable (for business), and should detail the impact as well as the likelihood of specific events (both favorable and unfavorable) occurring. In this plan it is important to include the responsibilities and time frames for managing such risks.

From a generic perspective, at least for business, the risk management plan should include the five main approaches to risk:

1. Remove the organization from the risk. For example, changing what the business does or withdrawing from a particular market.
2. Share or transfer the risk. For example, insuring or perhaps subcontracting work out.
3. Mitigate the risk. This would include taking actions to reduce the likelihood or impact of isolated factors.
4. Do nothing but ensure that there are contingency plans in place.
5. Accept the risk as being at a level of impact that can be "lived with." We all must be cognizant and accept that risk management is not about avoiding risk but identifying and managing it and taking chances on an informed basis. (This is intelligent strategic progression and it is most definitely a major contributor to overall success.)

To accomplish this plan, all parties involved (including senior management) must participate in an initial risk session, which should last about 4 to 8 hours. This time is quite negligible when one considers the future success of the organization and the consequences of not taking the time to evaluate the possible risks. In this meeting, all components of how the undertaking operates will be explored and risks identified and assessed in terms of likelihood of occurrence and impact. Key points are discussed, agreed on, and placed on a risk map before discussing an action plan, which must be regularly reviewed.

Risk Management and Business Continuity

Risk management is simply a practice of systematically selecting cost-effective approaches for minimizing the effect of threat realization to the organization. All risks can never be fully avoided or mitigated simply because of financial and practical limitations. Therefore, all organizations have to accept some level of residual risks (Mobley nd, Brown nd).

Whereas risk management tends to be preemptive, business continuity planning (BCP, an interdisciplinary peer mentoring methodology used to create and validate an exercised logistical plan for how an organization will recover and restore partially or completely interrupted critical function(s) within a predetermined time after a disaster or extended disruption) was invented to deal with the consequences of realized residual risks (Mobley nd).

The necessity to have BCP in place arises because even very unlikely events will occur if given enough time. Risk management and BCP are often mistakenly seen as rivals or overlapping practices. In fact, these processes are so tightly tied together that such separation seems artificial. For example, the risk management process creates important inputs for the BCP (assets, impact assessments, cost estimates, etc.). Risk management also proposes applicable controls for the observed risks. Therefore, risk management covers several areas that are vital for the BCP process. However, the BCP process goes beyond risk management's preemptive approach and moves on from the assumption that the disaster will realize at some point (risk assessment, risk management). In PM, risk management includes the following activities:

- Planning how risk management will be held in the particular project. Plan should include risk management tasks, responsibilities, activities, and budget.
- Assigning a risk officer—a team member other than a project manager who is responsible for foreseeing potential project problems. A typical characteristic of a risk officer is a healthy skepticism.
- Maintaining live project risk database. Each risk should have the following attributes: opening date, title, short description, probability, and importance. Optionally, a risk may have an assigned person responsible for its resolution and a date by which the risk must be resolved.
- Creating an anonymous risk reporting channel. Each team member should have possibility to report any risk foreseen in the project.
- Preparing mitigation plans for risks that are chosen to be mitigated. The purpose of the mitigation plan is to describe how this particular risk will be handled: what, when, by whom, and how it will be done to avoid it or minimize consequences if it becomes a liability.
- Summarizing planned and faced risks, effectiveness of mitigation activities, and effort spent for the risk management (Mobley nd).

RISK

Everyday life is full of physical risk. We cross the road, drive our cars, ride bikes through busy cities, climb ladders, light fires, wield sharp knives, and use power tools. Yet, in the developed world at least, society does its best to minimize these risks through legislation or public service advice. We have pelican crossings and roadside barriers, seat belts and side impact protection systems, bike lanes and helmets, fireguards and goggles, control of lethal weapons, and strict advice on the safe use of drills and chainsaws (Alderson 2008).

Though society tries to eliminate as much of this risk as possible, the urge remains—perhaps it is even strengthened—to seek thrills and put ourselves in potential physical danger. At the sanitized and consumerized end are scream-a-minute theme park rides, where families flock to feel the surge of adrenalin. Though these may satisfy a basic urge to scare oneself, they are very much part of our regulated and safety conscious society; the thrill is manufactured and almost always harmless.

Not so in the world of extreme sports, where people pit themselves against the elements and knowingly put themselves in risky, often life-threatening situations for pleasure. Not so in the world of business, where people can lose their investments and much more. The dangers in both sports and business are real and in some cases elemental and financially catastrophic: people die or lose everything. So why, in a world that tries to regulate out risk do some individuals return again and again to the mountains, the seas, the sky, and to risky investments to achieve what, to many, appears an entirely pointless outcome? The answer comes from those who indulge in adrenalin sports who often talk of living in that moment, of shutting out the world by complete concentration on the danger at hand. In the business world, it is the thrill of challenge and winning.

Danger is a concept that denotes a potential negative impact to an asset or some characteristic of value that may arise from some present process or future event. In everyday usage, *risk* is often used synonymously with the probability of a loss. In professional risk assessments, risk combines the probability of an event occurring with the impact that event would have and with its different circumstances. Risk is so important in everything that we do that we even have standards that define what risks are. For example, There are three main Australia and New Zealand standards that cover risk management: (1) AS4360:2004—Risk Management (2004), (2) HB436:2004—Risk Management Guidelines (2005), and (3) HB221:2004—Business Continuity Management (2004a).

Risk does not always only refer to the avoidance of negative outcomes. In game theory and finance, risk is only a measure of the variance of possible outcomes. Quality is an issue of Type I (alpha error) or Type II (beta error). Insurance is a classic example of an investment that reduces risk—the buyer pays a guaranteed amount and is protected from a potential large loss. Gambling is a risk-increasing investment, wherein money on hand is risked for a possible large return but also the possibility of losing it all. By this definition, purchasing a lottery ticket is an extremely risky investment (a high chance of no return but a small chance of a huge return), whereas putting money in a bank at a defined rate of interest is a risk-averse course of action (a guaranteed return of a small gain).

There are many definitions of risk depending on the specific application and situational contexts. Generally, risk is related to the expected losses that can be caused by a risky event and to the probability of this event. The worse the consequences of the loss are and the more likely the event, the worse the risk. Measuring risk is often difficult; rare failures can be hard to estimate, and loss of human life is generally considered irreplaceable. The engineering definition of risk is

Risk = (Probability of an accident) × (Losses per accident)

On the other hand, financial risk is often defined as the unexpected variability or volatility (*volatility* most frequently refers to the standard deviation of the change in value of a financial instrument) with a specific time horizon. It is often used to quantify the risk of the instrument over that time period. Volatility is typically expressed in annualized terms, and it may either be an absolute number ($$$) or a fraction of the initial value (*X* percent) of returns and thus includes both potential worse than expected as well as better than expected returns. References to negative risk below should be read as applying to positive impacts or opportunity (e.g., for loss read "loss or gain") unless the context precludes this.

In statistics, risk is often mapped to the probability of some event that is seen as undesirable. Usually the probability of that event and some assessment of its expected harm must be combined into a believable scenario (an outcome) that combines the set of risk, regret, and reward probabilities into an expected value. (In probability theory the expected value [or mathematical expectation] of a random variable is the sum of the probability of each possible outcome of the experiment multiplied by its payoff [value] for that outcome.)

Thus, in statistical decision theory, the risk function of an estimator $\delta(x)$ for a parameter θ, calculated from some observables x, is defined as the expectation value of the loss function (in statistics, decision theory, and economics, a *loss function* is a function that maps an event [technically an element of a sample space] onto a real number representing the economic cost or regret associated with the event; Frank 2008). Loss functions are typically expressed in monetary terms. For example,

$$\$ = \frac{\text{loss}}{\text{time period}}$$

Other measures of cost are possible; for example, mortality or morbidity in the field of public health or safety engineering (risk assessment). Loss functions are complementary to utility functions that represent benefit and satisfaction. Typically, for utility U the loss is represented by $k - U$, where k is some arbitrary constant) L,

$$R(\theta, \delta(x)) = \int L(\theta, \delta(x)) \times f(x|\theta)dx$$

where
   $\delta(x)$ = estimator
   θ = the parameter of the estimator

There are many informal methods used to assess or to measure risk, although it is not usually possible to directly measure risk. Formal methods measure the value at risk.

In scenario analysis (a process of analyzing possible future events by considering alternative possible outcomes [scenarios] the analysis is designed to allow improved decision making by allowing more complete consideration of outcomes and their implications. Risk is distinct from threat. A threat is a very low-probability but serious event—some analysts may be unable to assign a probability in a risk assessment because it has never occurred and there is no effective preventive measure (a step taken to reduce the probability or impact of a possible future event) available. The difference is most clearly illustrated by the precautionary principle, which seeks to reduce threat by requiring it to be reduced to a set of well-defined risks before an action, project, innovation, or experiment is allowed to proceed. In information security a risk is defined as a function of three variables:

1. The probability that there is a threat
2. The probability that there are any vulnerabilities
3. The potential impact

If any of these variables approaches zero, the overall risk approaches zero. Here we must also mention that the management of actuarial risk (as in insurance of all types) is called *risk management.*

Therefore, risk analysis can refer to (a) risk analysis (engineering), (b) probabilistic risk assessment (an engineering safety analysis), and (c) risk analysis (business). However, risk analysis is employed in its broadest sense to include two items.

1. *Risk assessment* involves identifying sources of potential harm, assessing the likelihood that harm will occur, and the consequences if harm does occur. Risk management also evaluates which risks identified in the risk assessment process require management and selects and implements the plans or actions that are required to ensure that those risks are controlled.
2. *Risk communication* involves an interactive dialogue between stakeholders and risk assessors and risk managers that actively informs the other processes.

Risk analysis, then, is equal to risk assessment plus risk management plus risk communication and it addresses at a minimum the following questions:

- How much risk is acceptable?
- Who benefits and who suffers the risks of the practice?
- Is the product really necessary?
- Are there less hazardous ways to satisfy human needs?
- Who should bear the burden of proof in decisions about technologies?
- Why does society assume that firms have a right to produce, use, and discharge toxic chemicals at all? (Thornton, 2000).

Risk management is the process of measuring (it involves estimating the ratio of the magnitude of a quantity to the magnitude of a unit of the same type [e.g., length, time, mass, etc.]; a measurement is the result of such a process, expressed as the product of a real number and a unit, where the real number is the estimated ratio), or assessing, risk and developing strategies to manage it. Strategies include transferring the risk to another party, avoiding the risk, reducing the negative effect of the risk, and accepting some or all of the consequences of a particular risk. Traditional risk management focuses on risks stemming from physical or legal causes (e.g., natural disasters or fires, accidents, death, and lawsuits). Financial risk management (the practice of creating value in a firm by using financial instruments to manage exposure to risk), on the other hand, focuses on risks that can be managed using traded financial instruments (Major nd).

In ideal risk management, a prioritization process is followed whereby the risks with the greatest loss and the greatest probability (extent to which something is likely to happen or be the case; generally it varies from one to zero) of occurring are handled first, and risks with lower probability of occurrence and lower loss are handled later. In practice, the process can be very difficult, and balancing between risks with a high probability of occurrence but lower loss vs. a risk with high loss but lower probability of occurrence can often be mishandled.

Intangible risk management identifies a new type of risk—a risk that has a 100 percent probability of occurring but is ignored by the organization due to a lack of identification ability. For example, knowledge risk occurs when deficient knowledge is applied.

Relationship risk occurs when collaboration ineffectiveness occurs. Process-engagement risk occurs when operational ineffectiveness occurs. These risks directly reduce the productivity of knowledge workers, decrease cost effectiveness, profitability, service, quality, reputation, brand value, and earnings quality. Intangible risk management allows risk management to create immediate value from the identification and reduction of risks that reduce productivity.

Risk management also faces difficulties allocating resources. This is the idea of opportunity cost. (In economics, opportunity cost, or economic cost, is the cost of something in terms of an opportunity forgone [and the benefits that could be received from that opportunity] or the most valuable forgone alternative [or highest-valued option forgone]; i.e., the second best alternative. Opportunity cost need not be assessed in monetary terms but rather can be assessed in terms of anything that is of value to the person or persons doing the assessing.) Resources spent on risk management could have been spent on more profitable activities. Again, ideal risk management minimizes spending while maximizing the reduction of the negative effects of risks (Risk Management). The steps in the risk management process are as follows:

1. Establish the context. Establishing the context involves:
   a. Planning the remainder of the process.
   b. Mapping out the following: the scope of the exercise, the identity and objectives of stakeholders, and the basis upon which risks will be evaluated.

c. Defining a framework for the process and an agenda for identification.

d. Developing an analysis of risk involved in the process.

2. Identify the risks. After establishing the context, the next step in the process of managing risk is to identify potential risks. Risks are about events that, when triggered, cause problems. Hence, risk identification can start with the source of problems or with the problem itself.

a. *Source analysis* risk sources may be internal or external to the system that is the target of risk management. Examples of risk sources are stakeholders of a project, employees of a company, or the weather over an airport.

b. *Problem analysis* risks are related to identified threats; for example, the threat of losing money, the threat of abuse of privacy information, or the threat of accidents and casualties. The threats may exist with various entities, such as shareholders, customers, and legislative bodies such as the government.

When either the source or problem is known, the events that a source may trigger or the events that can lead to a problem can be investigated. For example, stakeholders withdrawing during a project may endanger funding of the project; privacy information may be stolen by employees even within a closed network; lightning striking a Boeing 747 during takeoff may make all people onboard immediate casualties. The chosen method of identifying risks may depend on culture, industry practice, and compliance. The identification methods are formed by templates or the development of templates for identifying a source, problem, or event (see Risk Management Process (a and b); Common Risk). Common risk identification methods are as follows (Carothers nd):

- *Objectives-based risk* identification: Organizations and project teams have objectives. Any event that may endanger achieving an objective partly or completely is identified as risk.

- *Scenario-based risk* identification: In scenario analysis (a process of analyzing possible future events by considering alternative possible outcomes [scenarios]; the analysis is designed to allow improved decision making by allowing more complete consideration of outcomes and their implications) different scenarios are created. The scenarios may be the alternative ways to achieve an objective or an analysis of the interaction of forces in, for example, a market or battle. Any event that triggers an undesired scenario alternative is identified as risk.

- *Taxonomy-based risk* identification: The taxonomy in taxonomy-based risk identification is a breakdown of possible risk sources. Based on the taxonomy and knowledge of best practices, a questionnaire is compiled. The answers to the questions reveal risks (Crockford 1986).
- *Common risk checking:* In several industries, lists with known risks are available. Each risk in the list can be checked for application to a particular situation. An example of known risks in the software industry is the common vulnerability and exposures list found at http://cve.mitre.org.

Assessment of Risk(s)

Once risks have been identified, they must be assessed as to their potential severity of loss and the probability of occurrence. These quantities can be either simple to measure, in the case of the value of a lost building, or impossible to know for sure, in the case of the probability of an unlikely event occurring. Therefore, in the assessment process it is critical to make the best educated guesses possible in order to properly prioritize the implementation of the risk management plan. (The risk management plan describes how risk identification, qualitative and quantitative analysis, response planning, monitoring, and control will be structured and performed during the project life cycle. It defines roles and responsibilities for participants in the risk processes, the risk management activities that will be carried out, the schedule and budget for risk management activities, and any tools and techniques that will be used.)

The fundamental difficulty in risk assessment is determining the rate of occurrence, because statistical information is not available on all kinds of past incidents. Furthermore, evaluating the severity of the consequences (impact) is often quite difficult for immaterial assets. Asset valuation is another question that needs to be addressed. Thus, best educated opinions and available statistics are the primary sources of information. Nevertheless, risk assessment should produce such information for the management of the organization so that the primary risks are easy to understand and the risk management decisions may be prioritized. Thus, there have been several theories and attempts to quantify risks (Alderson 2008). Numerous different risk formulas exist, but perhaps the most widely accepted formula for risk quantification is

$$\text{Rate of occurrence} \times \text{Impact of the event} = \text{Risk}$$

This formula is imbedded in the failure mode and effect analysis (see Stamatis 2003a).

Research also has shown that the financial benefits of risk management are less dependent on the formula used but are more dependent on the frequency and how risk assessment is performed. After all, in business it is imperative to be able to present the findings of risk assessments in financial terms. Robert Courtney Jr., while working for IBM during the 1970s, proposed a formula for presenting risks in financial terms (Courtney 1970). The Courtney formula was accepted as the official risk analysis method for the U.S. governmental agencies. The formula proposes calculation of annualized loss expectancy (ALE) and compares the expected loss value to the security control implementation costs (cost–benefit analysis).

Potential Risk Treatments

Once risks have been identified and assessed, all techniques to manage the risk fall into one or more of these four major categories—remembered as the 4Ts (Dorfman 1997).

1. Tolerate (i.e., retention)
2. Treat (i.e., mitigation)
3. Terminate (i.e., elimination)
4. Transfer (i.e., buying insurance)

Ideal use of these strategies may not be possible. Some of them may involve trade-offs that are not acceptable to the organization or person making the risk management decisions.

Risk Avoidance

Risk avoidance includes not performing an activity that could carry risk. An example would be not buying a property. Property designates those things that are commonly recognized as being the possessions. Possession is having some degree of control over something else; for example, of a person or group. Important types of property include real property (land), personal property (other physical possessions), and intellectual property (rights over artistic creations, inventions, etc.). A right of ownership is associated with property that establishes the good as being "one's own thing" in relation to other individuals or groups, assuring the owner the right to dispense with the property in a manner in which he or she sees fit, whether to use or not use, exclude others from using, or to transfer ownership or business in order to avoid taking on the liability (a *liability* is anything that is a hindrance or puts individuals at a disadvantage) that comes with it. Another would be not flying in order to not take the risk that the airplane might be hijacked. Avoidance may seem the answer to all risks, but avoiding risks also means losing out on the potential gain that accepting (retaining) the risk may have allowed. Not entering a business to avoid the risk of loss also avoids the possibility of earning profits (Risk Management (c)).

Risk Reduction

Risk reduction involves methods that reduce the severity of the loss. Examples include sprinklers designed to put out a fire to reduce the risk of loss by fire. This method may cause a greater loss by water damage and therefore may not be suitable. Halon is a fire extinguishing system that, when used, does not leave a residue upon evaporation. It is considered a clean agent for eliminating fires. Halon is a liquefied, compressed gas that stops the spread of fire by chemically disrupting combustion. (Halon nd, Support nd). (Specifically, Halon can refer to (a) the haloalkanes [also known as halogenoalkanes], which are a group of chemical compounds, consisting of alkanes, such as methane or ethane, with one or more halogens linked, such as chlorine or fluorine, making them a type of organic halide; (b) Halon 1211 and Halon 1301, which are special-purpose fire-extinguishing agents that were banned by the Montreal Protocol [The Montreal Protocol on Substances That Deplete the Ozone Layer is an international treaty designed to protect the ozone layer by phasing out the production of a number of substances believed to be responsible for ozone depletion]; (c) Halon, which is an instrumental four-piece, hailing from Toledo, Ohio; (d) Halon, which is a Swedish firewall suppression system that may mitigate that risk, but the cost may be prohibitive as a strategy. A *strategy* is a long-term plan of action designed to achieve a particular goal. On the other hand, risk reduction is a strategy for minimizing risk. In other words, strategy is differentiated from tactics or immediate actions with resources at hand. Originally confined to military matters, the word has become commonly used in many disparate fields.)

Modern software development methodologies reduce risk by developing and delivering software incrementally. Early methodologies suffered from the fact that they only delivered software in the final phase of development; any problems encountered in earlier phases meant costly rework and often jeopardized the whole project. By developing in iterations, software projects can limit effort wasted to a single iteration. A current trend in software development, spearheaded by the extreme programming community, is to reduce the size of iterations to the smallest size possible; sometimes as little as one week is allocated to iteration. (Extreme programming, or XP, is a software engineering methodology and the most prominent of several agile software development methodologies. Like other agile methodologies, extreme programming differs from traditional methodologies primarily in placing a higher value on adaptability than on predictability.)

Risk Retention

Risk retention involves accepting the loss when it occurs. True self-insurance is a risk management method whereby an eligible risk is retained but a calculated amount of money is set aside to compensate for the potential future loss. The amount is calculated using actuarial and insurance information and the law of large numbers so that the amount set aside (similar to an insurance premium) is enough to cover the future uncertain loss. (The law of large numbers is a fundamental concept in statistics and probability that describes how the average of a randomly selected large sample from a population is likely to be close to the average of the whole population.) Self-insurance is similar to insurance in concept but involves either the payment of a self-insurance premium to a captive insurance company, cell captive or rent-a-captive insurer, or making an on-balance sheet provision and not paying a premium to an insurer at all. Risk retention is a viable strategy for small risks where the cost of insuring against the risk would be greater over time than the total losses sustained. All risks that are not avoided or transferred are retained by default. This includes risks that are so large or catastrophic that they either cannot be insured against or the premiums would be infeasible. War is an example, because most property and risks are not insured against war, so the loss attributed to war is retained by the insured. Also, any amount of potential loss (risk) over the amount insured is retained risk. This may also be acceptable if the chance of a very large loss is small or if the cost to insure for greater coverage amounts is so great it would hinder the goals of the organization too much.

Risk Transfer

Risk transfer means causing another party to accept the risk, typically by contract or by hedging. (In finance, a hedge is an investment that is taken out specifically to reduce or cancel out the risk in another investment. Hedging is a strategy designed to minimize exposure to an unwanted business risk, while still allowing the business to profit from an investment activity.) Insurance is one type of risk transfer that uses contracts. Other times it may involve contract language that transfers a risk to another party without the payment of an insurance premium. Liability among construction or other contractors is very often transferred this way. On the other hand, taking offsetting positions in derivatives (in finance, a *derivative* is a financial instrument derived from some other asset; rather than trade or exchange the asset itself, market participants enter into an agreement to exchange money, assets, or some other value at some future date based on the underlying asset) is typically how firms use hedging to financially manage risk. (Financial risk management is the practice of creating value. In general, the economic value of something is how much a product or service is worth to someone relative to other things [often measured in money].) It can be either an evaluation of what it could or should be worth or an explanation of its actual market value (price). Also, there are various ways to give those valuations or explanations. They are the subject of the theory of value in a firm by using financial instruments (Financial instruments are either a real or virtual document representing a legal agreement involving some sort of monetary value.) They can be categorized by form depending on whether they are cash instruments or derivative instruments (Risk Management (c)).

- *Cash instruments* are financial instruments whose value is determined directly by markets. They can be divided into securities, which are readily transferable, and other cash instruments such as loans and deposits, where both borrower and lender have to agree on a transfer.
- *Derivative instruments* are financial instruments that derive their value from some other financial instrument or variable. They can be divided into exchange traded derivatives and over-the-counter (OTC) derivatives.

Alternatively, they can be categorized by asset class depending on whether they are equity based (reflecting ownership of the issuing entity) or debt based (reflecting a loan the investor has made to the issuing entity). If it is debt, it can be further categorized into short term (less than one year) or long term.

Foreign exchange instruments and transactions are neither debt nor equity based and belong in their own category to manage exposure to risk. Similar to general risk management, financial risk management requires identifying the sources of risk, measuring risk, and formulating plans to address them. As a specialization of risk management, financial risk management focuses on when and how to hedge using financial instruments to manage costly exposures to risk.

Some ways of managing risk fall into multiple categories. Risk retention pools are technically retaining the risk for the group, but spreading it over the whole group involves transfer among individual members of the group. This is different from traditional insurance, in that no premium is exchanged between members of the group up front; instead, losses are assessed to all members of the group.

Planning for Risk

All projects should begin with some level of planning. Therefore, deciding on the combination of methods to be used for each risk is very important and circumstantial (specific to the project at hand). Each project should have its own risk management decision(s) and should be recorded and approved by the appropriate level of management. For example, a risk concerning the image of the organization should have top management behind it, whereas IT management would have the authority to decide on computer virus risks.

The risk management plan should propose applicable and effective security controls for managing the risks. For example, an observed high risk of computer viruses could be mitigated by acquiring and implementing the use of antivirus software. A good risk management plan should contain a schedule for control implementation and responsible persons for those actions. The risk management concept is old but is still not very effectively measured.

As a result of this vagueness in measurement, it is imperative not only that we implement appropriate and applicable risk methodologies but, just as important, review and evaluate the plan for our management risk plans. Therefore, we must follow all of the planned methods for mitigating the effect of the risks; for example, purchase insurance policies for the risks that will be transferred to an insurer, avoid all risks that can be avoided without sacrificing the entity's goals, reduce others, and retain the rest.

The assumption here is that the initial risk management plans will never be perfect. Practice, experience, and actual loss results will necessitate changes in the plan and contribute information to allow possible different decisions to be made in dealing with the risks being faced. Risk analysis results and management plans should be updated periodically. There are two primary reasons for this:

1. To evaluate whether the previously selected security controls are still applicable and effective.
2. To evaluate the possible risk level changes in the business environment. For example, information risks are a good example of rapidly changing business environment.

Limitations

If risks are improperly assessed and prioritized, time can be wasted in dealing with the risk of losses that are not likely to occur. Spending too much time assessing and managing unlikely risks can divert resources that could be used more profitably. Unlikely events do occur, but if the risk is unlikely enough to occur it may be better to simply retain the risk and deal with the result if the loss does in fact occur.

Prioritizing too highly the risk management processes could keep an organization from ever completing a project or even getting started. This is especially true if other work is suspended until the risk management process is considered complete. It is also important to keep in mind the distinction between risk and uncertainty. Risk can be measured by Impacts × Probability.

Areas of Risk Management

As applied to corporate finance, risk management is a technique for measuring, monitoring, and controlling the financial or operational risk on a firm's balance sheet. (Corporate finance is a specific area of finance which deals with the financial decisions that corporations make and the tools, as well as analysis used, to make these decisions. The primary goal of corporate finance is to enhance corporate value without taking excessive financial risks. The discipline may be divided among long-term and short-term decisions and techniques.) The Basel II (also called the New Accord; the correct full name is the International Convergence of Capital Measurement and Capital Standards—A Revised Framework) is the second Basel Accord and represents recommendations by bank supervisors and central bankers from the 13 countries making up the Basel Committee on Banking Supervision (BCBS) to revise the international standards for measuring the adequacy of a bank's capital. It was created to promote greater consistency in the way that banks and banking regulators approach risk management across national borders. The Bank for International Settlements (often confused with the BCBS; it supplies the secretariat for the BCBS but is not itself the BCBS) framework breaks risks into market risk (price risk), credit risk, and operational risk and also specifies methods for calculating capital requirements for each of these components (Hagar nd).

Enterprise Risk Management

In enterprise risk management, a risk is defined as a possible event or circumstance that can have negative influences on the enterprise in question. Its impact can be on the very existence, the resources (human and capital), the products and services, or the customers of the enterprise, as well as external impacts on society, markets, or the environment. In addition, every probable risk can have a preformulated plan to deal with its possible consequences (to ensure contingency if the risk becomes a liability).

From the information above and the average cost per employee over time, or cost accrual ratio (the cost accrual ratio for a business may be defined as the total average cost per person per unit time; e.g., average cost per day per person; it is only useful for risk assessment in small projects where average wages are roughly equal), a project manager can estimate

- The cost associated with the risk if it arises, estimated by multiplying employee costs per unit time by the estimated time lost. (Cost impact C, where C = cost accrual ratio × S [the cost accrual ratio for a business may be defined as the total average cost per person per unit time; e.g., average cost per day per person. It is only useful for risk assessment in small projects where average wages are roughly equal].)
- The probable increase in time associated with a risk (schedule variance due to risk, Rs where $Rs = P \times S$).

- Sorting on this value puts the highest risks to the schedule first. This is intended to cause the greatest risks to the project to be attempted first so that risk is minimized as quickly as possible.
- This is slightly misleading, because schedule variances with a large P and small S and vice versa are not equivalent (the risk of the RMS *Titanic* sinking vs. the passengers' meals being served at slightly the wrong time).
- The probable increase in cost associated with a risk (cost variance due to risk, Rc where $Rc = P \times C = P \times CAR \times S = P \times S \times CAR$).
  - Sorting on this value puts the highest risks to the budget first.
  - See concerns about schedule variance as this is a function of it, as illustrated in the Rc equation above.

Risk in a project or process can be due either to special cause variation or common cause variation and requires appropriate treatment; that is, to reiterate the concern about extreme cases not being equivalent in the list immediately above (Hajar nd).

PRINCE

An alternative to the classic approach of PM is the PRINCE model. PRINCE (PRojects IN Controlled Environments) is a structured method for effective PM. It is a de facto standard used extensively by the UK government and is widely recognized and used in the private sector, both in the UK and internationally. PRINCE, the method, is in the public domain, offering nonproprietorial best-practice guidance on PM. PRINCE is, however, a registered trademark of the Office of Government Commerce (OGC) and the copyright is retained by the Crown (PRINCE 2).

PRINCE was established in 1989 by the Central Computer and Telecommunications Agency (CCTA), since renamed the OGC. The method was originally based on PROMPT, a PM method created by Simpact Systems Ltd. in 1975. PROMPT was adopted by CCTA in 1979 as the standard to be used for all government information system projects. When PRINCE was launched in 1989, it effectively superseded PROMPT within government projects.

PRINCE provides a product-based start to the planning activity for any project. In other words, PRINCE is a generic, tailorable, simple-to-follow

PM method. It covers how to organize, manage, and control your projects. It is aimed at enabling you to successfully deliver the right products on time and within budget. As a project manager, you can apply the principles of PRINCE and the associated training to any type of project. It will help you to manage risk, control quality, and change effectively, as well as make the most of challenging situations and opportunities that arise within a project. This means that PRINCE adopts principles that help in the successful completion of a given project. The fundamental principles are as follows:

- A project is a finite process with a definite start and end.
- Projects always need to be managed in order to be successful.
- For genuine commitment to the project, all parties must be clear about why the project is needed, what it is intended to achieve, how the outcome is to be achieved, and what their responsibilities are in that achievement.

PRINCE also provides a planning framework that can be applied to any type of project. This involves the following:

- Establishing what products are needed.
- Determining the sequence in which each product should be produced.
- Defining the form and content of each product.
- Resolving what activities are necessary for their creation and delivery.

PRINCE2 is the second revision of PRINCE. It is a continuation of the process-based approach for PM providing an easily tailored and scaleable method for the management of all types of projects. Each process is defined with its key inputs and outputs together with the specific objectives to be achieved and activities to be carried out (PRINCE 2a, b, c, d). In a pictorial form it looks something like Figure 3.2.

Directing a Project

Directing a project (DP) runs from the startup of the project until its closure. This process is aimed at the project board. The project board manages by exception, monitors via reports, and controls through a number of decision points. However, it is imperative to recognize that this process does not cover the day-to-day activities of the project manager. The key processes for the project board break into four main areas.

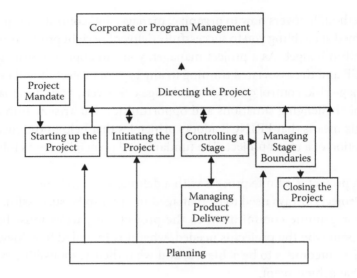

FIGURE 3.2
Overview of PRINCE2.

Initiation (Starting the Project Off on the Right Foot)

Starting up the project (SU) is the first process in PRINCE. It is a preproject process, designed to ensure that the prerequisites for initiating the project are in place. The process expects the existence of a project mandate, which defines in high-level terms the reason for the project and what outcome is sought. Starting up a project should be very short. The work of the process is built around the production of three elements:

1. Ensuring that the information required for the project team is available.
2. Designing and appointing the project management team.
3. Creating the initiation stage plan.

The objectives of initiating a project (IP) are to

- Agree on whether or not there is sufficient justification to proceed with the project.
- Establish a stable management basis on which to proceed.
- Document and confirm that an acceptable business case exists for the project.
- Ensure a firm and accepted foundation to the project prior to commencement of the work.

- Agree to the commitment of resources for the first stage of the project.
- Enable and encourage the project board to take ownership of the project.
- Provide the baseline for the decision-making processes required during the project's life.
- Ensure that the investment of time and effort required by the project is made wisely, taking into account the risks to the project.

Controlling a Stage

This process describes the monitoring and control activities of the project manager involved in ensuring that a stage stays on course and reacts to unexpected events. The process forms the core of the project manager's effort on the project, or the process that handles day-to-day management of the project. Throughout a stage there will be a cycle consisting of

- Authorizing work to be done
- Gathering progress information about that work
- Watching for changes
- Reviewing the situation
- Reporting
- Taking any necessary corrective action

This process covers these activities, together with the ongoing work of risk management and change control.

Managing Stage Boundaries

This process provides the project board with key decision points on whether to continue with the project or not. Furthermore, it demonstrates management's commitment by reviewing the resources used in the project and the future needs towards its completion. This review may also be in the form of an ad hoc direction (monitoring progress, providing advice and guidance, reacting to exception situations PRINCE 2e). The objectives of the stage boundaries (SB) process are to (PRINCE 2e)

- Assure the project board that all deliverables planned in the current stage plan have been completed as defined.

- Provide the information needed for the project board to assess the continuing viability of the project.
- Provide the project board with information needed to approve the current stage's completion and authorize the start of the next stage, together with its delegated tolerance level.
- Record any measurements or lessons that can help later stages of this project and/or other projects.

Closing a Project

The purpose of this process is to confirm the project outcome and execute a controlled close to the project. The process covers the project manager's work to wrap up the project either at its end or at premature close. Most of the work is to prepare input to the project board to obtain its confirmation that the project may close. The objectives of closing a project are, therefore, to

- Check the extent to which the objectives or aims set out in the project initiation document (PID) have been met.
- Confirm the extent of the fulfillment of the PID and the customer's satisfaction with the deliverables.
- Obtain formal acceptance of the deliverables.
- Determine to what extent all expected products have been handed over and accepted by the customer.
- Confirm that maintenance and operation arrangements are in place (where appropriate).
- Make any recommendations for follow-on actions.
- Capture lessons resulting from the project and complete the lessons learned report.
- Prepare an end project report.
- Notify the host organization of the intention to disband the project organization and resources.

Managing Product Delivery

The objective of this process is to ensure that planned products are created and delivered by

- Making certain that work on products allocated to the team is effectively authorized and all packages must meet the requirements upon acceptance.

- Ensuring that work conforms to the requirements of interfaces identified in the work package.
- Ensuring that the work is done.
- Assessing work progress and forecasts regularly.
- Ensuring that completed products meet quality criteria.
- Obtaining approval for the completed products.

Planning

Planning is a repeatable process and plays an important role in other processes, including the following:

- Planning an initiation stage
- Planning a project
- Planning a stage
- Producing an exception plan

The PRINCE2 system was first developed as the standard approach to IT PM for central government. Since then, the method has been enhanced to become a generic, best-practice approach suitable for the management of all types of projects and has a proven record outside both IT and government sectors. PRINCE2 has been widely adopted by both the public and private sectors and is now the UK's de facto standard for PM. There is also a rapidly growing international interest. PRINCE2 is designed to incorporate the requirements and experiences of existing users around the world. However, as powerful a methodology as it is, PRINCE2 does not cover all aspects of classic PM. Certain aspects of PM (such as leadership and people management skills, detailed coverage of PM tools and techniques) are well covered by other existing and proven methods and are therefore excluded from PRINCE2 (PRINCE 2 a, b, c, d, e).

In summary,

- PRINCE2 is a process approach to PM, fitting each process into a framework of essential components that need to be applied throughout the project.
- PRINCE2 helps you work out who should be involved in your projects, what they will be responsible for, and when they are likely to be needed. The set of processes and controls provided give you the structure that will support the life of the project and explains what information you should be gathering along the way.

- PRINCE2 method demonstrates how your project can be divided into manageable chunks or stages, allowing you to plan ahead more realistically and to call on your resources at the time they are most needed.
- PRINCE2 acts as a common language between all of customers, users, and suppliers, bringing these parties together on the project board. And although PRINCE2 does not include contract management as such, it provides the necessary controls and boundaries needed for everybody to work together within the limits of any relevant contracts. In addition, the project board provides support to the project manager in making key decisions.
- PRINCE2 allows the business to focus on doing the right projects, at the right time, for the right reasons, by making the start of a project and its continued existence dependent on a valid ongoing business case. It helps in establishing (a) strategic goals: where decisions are made, policies are set, and funding levels are set; (b) tactical goals: where decisions and policies are implemented within the allotted budgets; and (c) operational goals: where decisions translate into day-to-day actions and policies affect working practices within financial constraints.
- PRINCE2 is recognized as a world-class international product and is the standard method for PM, not least because it embodies many years of good practice in PM and provides a flexible and adaptable approach to suit all projects. It is a PM method designed to provide a framework covering the wide variety of disciplines and activities required within a project.
- The focus throughout PRINCE2 is on the business case, which describes the rationale and business justification for the project. The business case drives all of the PM processes, from initial project setup through to successful finish.
- PRINCE2 provides a mechanism to harness the skills and services of external suppliers, working alongside in-house resources, to enhance their ability to deliver successful projects with the given resources and enable the team to integrate and work together effectively on a project.

The key features and characteristics of PRINCE are the following:

- Its focus on business justification.

- Its defined organization structure for the project management team.
- Its product-based planning approach.
- Its emphasis on dividing the project into manageable and controllable stages.
- Its flexibility to be applied at a level appropriate to the project.
- Its finite and defined life cycle.
- Its defined and measurable business products.
- Its corresponding set of activities to achieve the business products.
- Its defined amount of resources.
- Its organization structure, with defined responsibilities, to manage the project.

BENEFITS OF USING PRINCE

PRINCE provides benefits to the managers and directors of a project and to an organization, through the controllable use of resources and the ability to manage business and project risk more effectively.

PRINCE embodies established and proven best practice in PM. It is widely recognized and understood, providing a common language for all participants in a project. PRINCE encourages formal recognition of responsibilities within a project and focuses on what a project is to deliver, why, when, and for whom. PRINCE provides projects with

- A controlled and organized start, middle, and end.
- Regular reviews of progress against plan and against the business case flexible decision points.
- Automatic management control of any deviations from the plan.
- The involvement of management and stakeholders at the right time and place during the project.
- Good communication channels between the project, PM, and the rest of the organization.

Those who will be directly involved with using the results of a project are able to

- Participate in all of the decision making on a project.

- Be fully involved in day-to-day progress, if desired.
- Provide quality checks throughout the project and ensure that their requirements are being adequately satisfied.

For senior management, PRINCE uses the management by exception concept. That means that management is kept fully informed of the project status without having to attend regular, time-consuming meetings. Typically, they are able to

- Establish terms of reference as a prerequisite to the start of a project.
- Use a defined structure for delegation, authority, and communication.
- Divide the project into manageable stages for more accurate planning.
- Ensure that resource commitment from management is part of any approval to proceed.
- Provide regular but brief management reports.
- Keep meetings with management and stakeholders to a minimum but at the vital points in the project.

The PRINCE2 method is documented in the OGC publication "Managing Successful Projects with PRINCE2" (Lea 2007), readily available from the official publisher, The Stationary Office (TSO). This core book is supported by a number of complementary publications that add guidance on how to tailor the method, how to manage your people in a PRINCE project, and outlining the benefits to your business of adopting a PRINCE2 approach.

ERROR ANALYSIS

As we get under way in the twenty-first century, one of the essential things that we all have to be concerned with is the idea of error—so much so that we have to make an extra effort to understand it.

A measurement may be made of a quantity that has an accepted value that can be looked up in a handbook (e.g., the density of a particular element). The difference between the measurement and the accepted value is not what is meant by error. Such accepted values are not "right" answers.

They are just measurements made by other people that have errors associated with them as well. Nor does *error* mean *blunder*. Reading a scale backwards, misunderstanding what you are doing, or inappropriate sequence in a particular process are blunders that can be caught and should simply be corrected and disregarded.

Obviously, it cannot be determined exactly how far off a measurement is; if this could be done, it would be possible to just give a more accurate, corrected value. Error, then, has to do with uncertainty in measurements that nothing can be done about. If a measurement is repeated, the values obtained will differ and none of the results can be preferred over the others. Although it is not possible to do anything about such error, it can be characterized. For instance, the repeated measurements may cluster tightly together or they may spread widely. This pattern can be analyzed systematically.

In the world of quality, with measurement, project execution, and many other domains, we are concerned with error. Error maybe the result of data definition, data identification, data collection, data analysis, and data interpretation. Generally, this error is called the *experimental error*. Error may be also generated due to human performance as well as the instrument/equipment that is used. Let us look at these distinct categories separately.

Experimental Error

The reason for having data is to draw sound conclusions and project those conclusions to a given situation. For anyone to draw sound conclusions it is crucial to understand that *all* measurements are subject to uncertainties. It is never possible to measure anything exactly. It is good, of course, to make the error as small as possible, but it is always there. In order to draw valid conclusions, the error must be indicated and dealt with properly. For example, take the measurement size of a person's shoe. Assuming that the size is 10½ inches, how accurate is that size?

Well, the shoe size depends on whether the person has curved toes or not; whether the foot is loose or stiff; whether the person wears thin, thick, or no socks; and so on. All of these characteristics play a major role in selecting the right shoe size as well as considering whether the person wants the shoes for walking, running, dress-up, or a casual purpose. These inaccuracies could all be called errors of definition. A quantity such as a shoe size is not exactly defined without specifying many other circumstances. Even if you could precisely specify the circumstances, your result would still have an error associated with it, depending on where the shoe was made; the scale you are using is of limited accuracy, depending on its resolution and so on.

If the result of a measurement is to have meaning, it cannot consist of the measured value alone. An indication of how accurate the result is must be included also. Indeed, typically more effort is required to determine the error or uncertainty in a measurement than to perform the measurement itself. Therefore, the result of any measurement has two essential components: (1) a numerical value (in a specified system of units) giving the best estimate possible of the quantity measured, and (2) the degree of uncertainty associated with this estimated value. Generally speaking, these measurements are shown as $\bar{X} \pm \sigma$ (the average value plus or minus the standard deviation) or the standard error of the mean (SEM). SEM is usually estimated by the sample estimate of the population standard deviation (s) divided by the square root of the sample size (assuming statistical independence of the values in the sample). The calculation for the SEM is

$$SE_{\bar{x}} = \frac{s}{\sqrt{n}}$$

where

s = the sample standard deviation (i.e., the sample-based estimate of the standard deviation of the population), and

n = the size (number of items) of the sample.

An example of a typical numerical measurement with an associated error estimate may be shown as 15.3 ± 0.1 cm.

Another concern with the error analysis is the issue of significant figures. The significant figures of a (measured or calculated) quantity are the meaningful digits in it. There are conventions that you should learn and follow for how to express numbers to properly indicate their significant figures. Any digit that is not zero is significant. So, 369 has three significant figures and 1.947 has four significant figures. Zeros between non-zero digits are significant. So 4023 has four significant figures. (Special note: zeros to the left of the first non-zero digit are not significant. Thus, 0.000021 has only two significant figures. This is more easily seen if it is written in scientific notation, which is 2.1×10^{-5}. For numbers with decimal points, zeros to the right of a non-zero digit are significant. So, 3.00 has three significant figures and 0.010 has two significant figures. For this reason it is important to keep the trailing zeros to indicate the actual number of significant figures.)

For numbers without decimal points, trailing zeros may or may not be significant. So, 100 indicates only one significant figure. To indicate that the trailing zeros are significant, a decimal point must be added. For example, "100." has three significant figures, and 1×10^2 has one significant figure. Exact numbers have an infinite number of significant digits. For example, if there are two apples in a bowl, then the number of apples is 2.000. Defined numbers are also like this. For example, the number of centimeters per inch (2.54) has an infinite number of significant digits, as does the speed of light (299,792,458 m/s).

There are also specific rules for how to consistently express the uncertainty associated with a number. In general, the last significant figure in any result should be of the same order of magnitude (i.e., in the same decimal position) as the uncertainty. Also, the uncertainty should be rounded to one or two significant figures. Always work out the uncertainty after finding the number of significant figures for the actual measurement.

For example, correct measurements are 1.32 ± 0.01; 19.1 ± 1.1; 8 ± 1.

On the other hand, the following numbers are all incorrect. 10.82 ± 0.021235 is wrong but 10.82 ± 0.02 is fine; 15.0 ± 1 is wrong but 15.0 ± 1.0 is fine; 2 ± 0.5 is wrong but 2.0 ± 0.5 is fine.

In practice, when doing mathematical calculations, it is a good idea to keep one more digit than is significant to reduce rounding errors. But in the end, the answer must be expressed with only the proper number of significant figures. After addition or subtraction, the result is significant only to the place determined by the largest last significant place in the original numbers. For example, $61.245 + 1.1 = 62.345$ should be rounded to get 62.3 (the tenths place is the last significant place in 1.1). After multiplication or division, the number of significant figures in the result is determined by the original number with the smallest number of significant figures. For example, $2.32 \times 5.2064 = 12.078848$ should be rounded off to 12.1 (three significant figures like 2.32). The exception to these rules is in calculating any reliability, in which case we use the entire number with all decimals. Significant figures generally do not apply. For a good discussion on significant figures, the reader is encouraged to refer to any basic math textbook.

CLASSIFICATION OF ERROR

Generally, errors can be divided into two broad and rough but useful classes: systematic and random. Systematic errors are errors that tend to shift all measurements in a systematic way so their mean value is displaced. This may be due to such things as incorrect calibration of equipment, consistently improper use of equipment, or failure to properly account for some effect. In a sense, a systematic error is rather like a constant being added (blunder), and large systematic errors can and must be eliminated in a good experiment. But small systematic errors will always be present. For instance, no instrument can ever be calibrated perfectly.

Other sources of systematic errors are external effects that can change the results of the experiment but for which the corrections are not well known. In a design of experiment, for example, several independent confirmations of experimental results are often required (especially using different techniques) because different apparatus at different places may be affected by different systematic effects. Aside from making resolution mistakes (such as thinking one is using the ×10 scale and actually using the ×100 scale), experiments sometimes yield results that may be far outside the quoted errors because of systematic effects not accounted for.

Random errors, on the other hand, are errors that fluctuate from one measurement to the next. They yield results distributed about some mean value. They can occur for a variety of reasons. They may occur due to lack of sensitivity (an instrument may not be able to respond to or indicate a sufficiently small change, or the observer may not be able to discern it). They may occur due to noise. There may be extraneous disturbances that cannot be taken into account. They may be due to imprecise definition. They may also occur due to statistical processes such as the roll of dice.

Random errors displace measurements in an arbitrary direction, whereas systematic errors displace measurements in a single direction. Some systematic error can be substantially eliminated (or properly taken into account). Random errors are unavoidable and must be lived with. Many times you will find results quoted with two errors. The first error quoted is usually the random error, and the second is called the *systematic error*. If only one error is quoted, then the errors from all sources are added together. Generally speaking, systematic errors are associated with assignable causes, and random errors are associated with inherent variation. In the first case, the operator may in fact be able to correct the problem; in the second, either management or new technology may improve the process. To identify, track, and separate each of the errors, the most common methodology is to use statistical process control (SPC).

The second error that we may encounter is in the area of *measurement*. It has been said that "if you cannot measure it, you cannot improve it (Kelvin nd)." Because quality initiatives are always focused on improvement, it is imperative that we measure everything that we do. Obviously, some measurements are more important than others and, as such, we need to know what we measure and how accurate we are in our measurements. In essence, a measurement system is the collection of operations, procedures, gage, software, and personnel used to assign a number to the characteristic being measured. Therefore, measurement system analysis (MSA) quantifies major sources of process variation that come to us from

- Materials
- Methods
- Machines
- People
- Environment
- Measures (the most neglected item)

Each one of these components of measurement error can contribute to variation, causing wrong decisions to be made. For the measurement error to be accurate, we must view it as a process and a system; it must explain part and gage variation; and it also must identify its effect on quality and identify measurement tools and their use (DaimlerChrysler Corporation, Ford Motor Corporation, and General Motor Corporation 2002). However, as a scientific and objective method of analyzing the validity of a measurement system, it has become a tool that quantifies

- Repeatability—sometimes called *equipment variation*
- Reproducibility—operator variation
- The total variation of a measurement system

This quantification of variation can be characterized by

- Location of the process
  - Accuracy/bias

- Stability (consistency)
- Linearity
- Width or spread (precision)
  - Repeatability
  - Reproducibility
- Resolution/discrimination

At this point we must emphasize that

- MSA is NOT just calibration
- MSA is NOT just gage repeatability and reproducibility (gage R&R)
- MSA means walking the process (see the gauging process, observe destructive testing, and reduce the error in the process)
- MSA means having specific operational definitions

Because measurement is a system, it also has a process that has to be followed. The process is divided into three steps:

1. The before stage, in which you
   a. Make adjustments
   b. Implement solutions
   c. Run an experiment
2. Statistical analysis: The actual process used to obtain measurements, which may be summarized as follows:
   a. Provide information about what is happening in the process that produced the part
   b. Good measurement data = Good decisions
   c. Ensure accuracy and precision
   d. Eliminate false readings that indicate something is wrong in the process
   e. Reduce error in calling good parts bad (Type I error or producer risk)
   f. Reduce error in calling bad parts good (Type II error or consumer risk)
3. The after stage, in which you
   a. Validate your measurement system or measurement process (use analysis of variance [ANOVA] method)
   b. Validate data and data collection systems

The effect of this process is twofold:

1. To understand whether a decision
    a. Meets standards and specifications.
    b. Is detection/reaction oriented.
    c. Provides short-term results.
2. Stimulate continuous improvement:
    a. Where to improve?
    b. How much to improve?
    c. Is improvement cost-effective?
    d. Is it prevention oriented?
    e. Has Long-term strategy been planned?

When the three stages are completed, one may characterize successful MSA as a system to

- Define and validate the measurement process.
- Identify known elements of the measurement process (operators, gages, standard operating procedures [SOP], setup, etc.).
- Clarify the purpose and strategy for evaluation.
- Set acceptance criteria—operational definitions.
- Implement preventive/corrective action procedures.
- Establish ongoing assessment criteria and schedules.

DEFINITIONS IN MSA

As in any methodology, there are several key definitions one must be aware of in MSA. These will be discussed next.

Resolution/Discrimination

Resolution/discrimination is the ability to measure small differences in an output. Unless the appropriate resolution is used, the inevitable will occur. Inadequate measurement resolution may be because the measurement graduations are too large to detect variation present. Use the 10-bucket rule. This means that the tolerance range of your process divided by 10 should be the approximate resolution of your measurement system. Another way of looking at this is that the increments in the measurement system should be at least one-tenth the product specification or process variation (depending on how the gage is being used). On the other hand:

- If the team is fixing the process, use process variation.
- If checking for conforming/nonconforming parts, use product specification.

Poor resolution is a common issue and, as such, it must be viewed as a simplest measurement system problem. The impact is rarely recognized and/or addressed. However, it is easily detected; no special studies are necessary; and no known standards are needed. If a resolution problem is detected, the following actions may be taken:

- Measure to as many decimal places as possible.
- Use a device that can measure smaller units.
- Live with it but document that the problem exists.
- Larger sample size may overcome problem.
- Priorities may need to involve other considerations:
  - Engineering tolerance
  - Process capability
  - Cost and difficulty in replacing device

Accuracy and Bias

Measurements are shifted from true value. Bias is the difference between the observed average value of measurements and the master value. If accuracy or a bias problem is detected, the following actions may be taken:

- Calibrate when needed/scheduled.
- Use operating instructions.
- Review operating specifications
- Review software logic (drill-downs).
- Create operational definitions.

Linearity

Measurement is not consistent across the range of the gage. If linearity is suspect, the following actions may be taken:

- Use only in restricted range (consider where the specification limits are).
- Rebuild.
- Use with correction factor/table/curve.
- Sophisticated study is required and will not be discussed in this book.

Stability

Measurement drifts. Measurements remain constant and predictable over time. This can be evaluated by using control charts. *Stability* refers to both mean and standard deviation; it means no drifts, sudden shifts, cycles, etc. If stability is not present, the following actions may be taken:

- Change/adjust components.
- Establish "life" timeframe.
- Use control charts.
- Use/update current SOP.

Precision

Precision is a combination of repeatability and reproducibility. A measurement with relatively small indeterminate error is said to have high precision. A measurement with small indeterminate error and small determinate error is said to have high accuracy. Precision does not necessarily imply accuracy. A precise measurement may be inaccurate if it has a determinate error.

Repeatability

Repeatability is variation that occurs when repeated measurements are made of the same item under absolutely identical conditions. That means that we have the same

- Operator
- Setup (usually changes between measurements)
- Units
- Environmental conditions
- Short-term or within-operator variation

If repeatability is suspect, the following actions may be taken:

- Repair, replace, or adjust equipment.
- Review and if necessary change or update the SOP—Standard Operating Procedure (i.e., clamping sequence, method, force).
- Gage maintenance may be needed.

Reproducibility

Reproducibility is the variation that results when different conditions are used to make the measurements. That means that situations are different, such as:

- Operators (most common)
- Setups
- Test units (end-of-line testers)
- Environmental conditions
- Locations
- Variation between different operators when measuring the same part
- Long-term or between-operator variation

If reproducibility is suspect, the following actions may be taken:

- Operators need more training in gage usage and data collection.
- Standardized work through an SOP.
- The gathering-data device graduations, part location points, and data collection locations all need to be more clearly defined and identified.
- Create better operational definitions.

When we are involved with gages, we must also be very cognizant of their care. Gages are sensitive instruments whose accuracy is adversely affected by abuse. All users of gages must properly use and care for those gages. When an instrument is inadvertently mishandled, it must be turned in to the calibration laboratory for checking and recalibration. Some basic guidelines for care and handling are as follows:

- Use the instrument only for the purposes for which it is intended.
- Never use any manipulative device or force the instrument.
- Clean and remove burrs from the piece or product prior to use.
- Never expose an instrument directly to heat.
- Avoid exposure to dust, moisture, and grease.
- Lightly oil instrument and store in a protected case or cabinet.
- Return all instruments to storage case or cabinet after use.

GAGE R&R—ATTRIBUTE MSA

The general process of the attribute-data MSA follows a six-step approach:

Step 1: Select a minimum of 30 parts from the process (recommended 50–100).
- 30–50 percent of the parts in your study should have defects.
- 50–70 percent of the parts should be defect free.
- If possible, select some borderline (or marginal) good and bad samples.

Step 2: Identify the appraisers, who should be qualified.

Step 3: Have each appraiser, independently and in random order, assess these parts and determine whether or not they pass or fail (must have documented operational definitions).

Step 4: Enter the data into Excel or MINITAB to report the effectiveness of the attribute measurement system.

Step 5: Document the results. Implement appropriate actions to fix the inspection process if necessary. Goal is to have 100 percent match within and between operators to the known or correct attribute.

Step 6: Rerun the study to verify the fix.

Note: A 30-piece sample will yield an estimate of appraiser efficiency and capability that has a fair amount of uncertainty. Typically a larger sample is not needed when the appraisal process is obviously ineffective. The Excel spreadsheet can handle up to 100 samples.

GAGE R&R VARIABLE-DATA MSA

The general process of the variable-data MSA follows six steps. The concern for all steps is a basic question: What would your long-term gage plan be? The answer to this question is the very essence of MSA, because it focuses on the integrity of the data and the ultimate decision that is made based on that data. The steps are as follows:

1. Conduct initial gage calibration (or verification).
2. Perform trials and data collection.
3. Obtain statistics via MINITAB.

4. Analyze and interpret ANOVA results.
5. Check gage R&R and graphical output.
6. Conduct ongoing evaluation and continuous improvement actions.

To conduct a complete gage R&R for variable data, within the overall steps there are three stages with specific requirements that must be completed: (a) preparation, (b) conducting the MSA, and (c) analysis of the data.

The preparation stage covers the following items:

Step 1: Establish the need for the study and information desired from it before any data are collected (percentage study variation, percentage tolerance).

Step 2: Determine the approach to be used and operational definitions.

Step 3: Determine the number of operators, number of sample parts, and the number of repeat readings in advance.
  • Criticality of the dimension/function
  • Part configuration
  • Level of confidence and precision that is desirable for the gage system error estimate

Step 4: Whenever possible, select the operators who normally operate the gage.

Step 5: Choose sample parts to represent the entire range of variability of the process (not tolerance; do not manipulate or doctor up parts).

Step 6: Ensure that the gage has graduations that are at least one-tenth of the tolerance or process variation of the characteristic being measured (depends on how the gage is being used—to check for good/bad parts or for controlling the process).

Step 7: Ensure that the measuring method and gage are both measuring the desired characteristic.

The second stage of conducting the MSA covers the following items:

Step 1: Take all measurements in a random order to ensure that any drift or changes that occur will be spread randomly throughout the study.

Step 2: Estimate all readings to the nearest number possible. At a minimum, all readings should be taken at half of the smallest gradation.

Step 3: If operator calibration is suspected to be a major influence on variation, the gage should be recalibrated by the operator before each trial.

Step 4: The study should be observed by someone who understands the importance of the precautions required to conduct a reliable study.

The mechanics of conducting the MSA should be based on some fundamental principles of data collection, such as:

- Generally two to three operators.
- Generally five to ten process outputs to measure.
- Each process output is measured two to three times (replicated) by each operator.
- Multiple opportunity measurement (crossed gage R&R). Based on this, (a) the same part can be measured several times, and (b) error or measurement can be estimated.
- Single opportunity measurement (nested gage R&R).
- Characteristics of the item being tested or the components of the measurement process are no longer the same as when the test began. Typical issues here are as follows:
  - Destructive testing
  - Hardness testing
  - Aging
  - Chemical analysis
- Need to collect a batch of "like" parts (minimize within the variation) for the measurement system study.

Using MINITAB, the process is as follows:

Step 1: Collect five to ten samples that represent the full range of long-term process variation. In addition, identify the operators who use this instrument daily.

Step 2: Calibrate the gage or verify that the last calibration date is valid.

Step 3: Setup the MINITAB data collection sheet for the R&R study; column headings: Part ID, operator, trial, measurement(s); Calc > Make Patterned Data > Simple Set of Numbers (for each input).

Step 4: Ask the first operator to measure all the samples once in random order. Blind sampling, in which the operator does not know the identity of each part, should be used to reduce human bias.

Step 5: Have the second operator measure all of the samples once in random order and continue until all operators have measured the samples once (this is trial 1).

Step 6: Randomize the parts again and repeat steps 4 and 5 for the required number of trials.

Step 7: Enter the data and tolerance information into MINITAB. Then go to:
- Stat > Quality Tools > Gage R&R Study
- Stat > Quality Tools > Gage Run Chart

Step 8: Analyze the results by assessing the quality of the measurement system based on the guidelines on pages 125 through 139. Determine follow-up actions.

The third stage of analysis maybe conducted in three different ways: (1) short- or long-form analysis as shown in the AIAG publication (DaimlerChrysler Corporation, Ford Motor Corporation, and General Motor Corporation 2002); (2) the graphical approach of the six panels as calculated with MINITAB, and (3) the ANOVA approach. In this section we will address the graphical approach, which is the simplest and most convenient, and the summary analysis of the ANOVA process. For the first one, the reader is encouraged to see AIAG (2002) and Stamatis (1997). The graphical approach is based on six charts, commonly known as the *six panels*. They are shown in Figure 3.3.

The pictorial analysis of these charts is very simple but very powerful in identifying concerns with the MSA. Typical things that we look for in these charts are as follows:

- Total gage R&R <30 percent (percentage study variation), and part-to-part to be the biggest contributor (upper left chart). Tall bar charts: Distinguishes the components of variation in percentages. Repeatability, reproducibility, and parts (want low gage R&R, high part-to-part variation).
- R-chart must be in-control. An out-of-control range chart indicates poor repeatability. Want to see five or more levels (look across the points) of range within the control limits (middle left). R-chart: Helps identify unusual measurements. Repeatability/resolution (no outliers permitted).
- Xbar chart must be 50 percent out-of-control or more (indicating that the measurement system can tell a good part from a bad part) and similar patterns between operators (lower left). Xbar chart: Shows sampled process output variety. Reproducibility/sensitivity (want similar patterns for each operator).

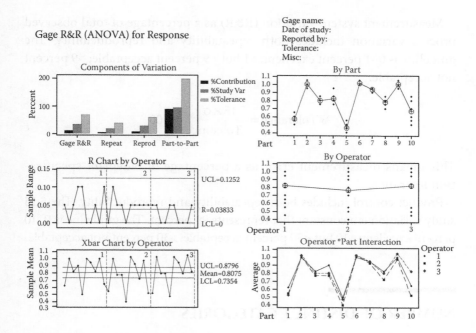

Gage R&R (ANOVA) for Response

Gage name:
Date of study:
Reported by:
Tolerance:
Misc:

FIGURE 3.3
Six panels for gage R&R.

- Want spread of the 10 MSA parts to represent the actual process variation (upper right).
- Want the operator means to be the same (straight line; middle right).
- Want lines to be parallel and close to each other. This indicates good reproducibility (lower right).

MEASURING MSA WITH INDICES

If the gage does not pass the percentage study variation, it cannot perform the job of process control (it will not be able to adequately distinguish one part from another). Measurement system standard deviation (R&R) as a percentage of total observed process standard deviation includes both repeatability and reproducibility. The guideline is 0–10 percent excellent; >10 but <30 percent acceptable; >30 percent not acceptable.

Same as percentage study variation, except a ratio of variances vs. standard deviations.

Measurement system variation (R&R) as a percentage of total observed process variation includes both repeatability and reproducibility. The guideline is 0–1 percent excellent; >1 but <9 percent acceptable; >9 percent not acceptable.

$$\% \text{ Tolerance} = \frac{5.15 \times \sigma_{R\&R}}{\text{Tolerance}} \times 100$$

This means measurement error as a percentage of blueprint specification tolerance.

Product control includes both repeatability and reproducibility. The 5.15 study variation is a constant that represents 99 percent. The guideline is 0–10 percent excellent; >10 but <30 percent acceptable; >30 percent not acceptable.

NUMBER OF DISTINCT CATEGORIES

The number of distinct categories is something that many people find confusing. However, it is a simple concept that is used to establish whether the measurement system has good or poor discrimination (resolution). The rule of thumb is for the increments of measure to be one-tenth of the process variation. It is used in a production environment when multiple parts can be gathered that represent the process, allowing part-to-part variation to be calculated. It can be calculated as follows.

In essence, the number of distinct categories is the number of divisions that the measurement system can accurately measure across the process variation (this is effective resolution, not gage resolution). Another way of thinking about it is how well a measurement process can detect process output variation, process shifts, and improvement. A typical guideline is 0–4 questionable; 5–10 acceptable; >5 excellent. If the categories are less than five, that indicates attribute conditions.

WHAT TO LOOK FOR IN HIGH GAGE R&R ERROR

The function of any MSA is to make sure that the data have integrity and the decisions we make based on that data are sound. So, when we are all done with the analysis, we should make sure that key measurement system evaluation questions have been addressed. These are as follows:

- Is there a written inspection/measurement procedure?
- Is there a detailed process map developed?
- Is there a specific measuring system and setup defined?
- Is there a trained or certified operator(s)?
- Is there an instrument calibration performed in a timely manner?
- Is there a record of tracking accuracy?
- Is there a tracking percentage of R&R?
- Is there a definition of an effective resolution?
- Is there a correlation with supplier or customer where appropriate?

Obviously, in any gage R&R study we want all four indices (bias, linearity, accuracy, stability) to be within the excellent or acceptable ranges and the graphical six panel to be OK. If not, we look at least at the following four areas:

1. Repeatability issues (within operator)
   a. No gauging instructions or operational definitions
   b. Lack of skill in reading vernier scales
   c. Clamping sequence differs from trial to trial
   d. Loose gage features/gage not rigid enough
   e. Measuring location on part differs from trial to trial
   f. Semidestructive testing changes the part (squeezing soft parts)
   g. Not enough gage resolution
   h. Rounding away (up and down) gage resolution
   i. Instrument needs maintenance
2. Reproducibility issues (between operators)
   a. No gauging instructions or operational definitions
   b. Differences in skill in reading vernier scales
   c. Clamping sequence differs from operator to operator
   d. Different (operator-owned?) measurement devices
   e. Ergonomics: operator size, height, strength, etc.
   f. Different rounding methods, operator to operator
   g. Temperature of gage over time; that is, expansion
3. Resolution: Rules for detecting inadequate measurement units.

    a. The measurement graduations borders may be too large when there are less than five possible values within the control limits on the range chart.

    b. Four values within the limits will be evidence of inadequate measurement units (one, two, or three possible values will result in substantial distortion).

    c. Also, beware if more than 25 percent of the ranges are zero.

4. Calibration steps: Determine whether the measurement system needs to be recalibrated.

    a. Determine the minimum number of measurements needed to make this decision (10 times).

    b. Take data and make decision. (If yes, recalibrate system. Why don't we just recalibrate? Because if we do so with no other changes, we are going to have the same results: Nothing changed! On the other hand, if we take new data and proceed with analysis we may find a different result.)

      i. Normal variation causes the measurement to be slightly different each time it is used.

     ii. Recalibration should be done only when the measurements are off by more than the normal variation.

    iii. Recalibrating a system when it is not needed can increase the variability in the measurements.

ANOVA

ANOVA is a statistical technique used to test the null hypothesis that several population means are equal. It is called ANOVA because it examines the variability in the sample and, based on that variability, it determines whether there is reason to believe that the population means are unequal. It also addresses the contribution of interaction between groups in the sample (Stamatis 2003b, 2003c).

In this section we are not going to give the statistical procedure and rationale of ANOVA. However, we are going to address the issue of ANOVA as it relates to MSA using MINITAB. The reader is encouraged to see the references at the end of this chapter and other statistical books to learn more about this topic.

Once you open MINITAB, you enter the data into the columns of the worksheet and then you go in the "Stat" command in the option bar. You click ANOVA and proceed in the session window (click on scroll icon), find the ANOVA of interest, and proceed with the props. When you are done, MINITAB will provide you with an ANOVA table.

Find the *P* (*p*-value) associated with "Parts"—it should be less than 0.05, meaning that the parts are a significant source of variation. They should be. If not, the parts may not be representative of the process or the measurement system is inadequate to tell the difference between parts. All other *p*s should be greater than 0.05. Look in the "Total Gage R&R" row under the headings "% Study Var" and "% Tolerance." Both should be less than 10 percent to be good; less than 30 percent may be acceptable. An acceptable "% Study Var" indicates that the measurement system is good enough for use in improving the process. An acceptable "% Tolerance" indicates that the measurement system is good enough to tell good parts from bad parts; that is, sorting.

If one or both of the values is unacceptable, look at the amount of each that is due to repeatability and reproducibility. If the percentage due to repeatability is high, this means that the same operator is not measuring the same part in the same way each time. If the percentage due to reproducibility is high, this means that different operators are, on average, measuring the same parts differently. Find out why!

Looking at a typical ANOVA table, you will see the column headings with the characteristics of the study followed by a table of numbers. That setup will look something like Table 3.2.

RULES OF THE MINITAB OUTPUT

In addition to the table, one may be interested in seeing the six-panel output that we discussed earlier (Figure 3.3). Here we give an overview of the graphical output.

1. *Components of Variation:* This chart summarizes the overall results from the session window. Part-to-part variation should account for almost all of the variation observed.

TABLE 3.2

Gage R&R Study—ANOVA Method

Gage name: EN/FN specific gage

Date of study: 10/22/11

Two-Way ANOVA Table with Interaction

Source	DF	SS	MS	F	P	
Parts	9	0.302462	0.0336069	179.799	0.00000	← Should be < 0.05
Operators	2	0.010747	0.0053733	28.748	0.00000	← Operators are significant
Operators × Parts	18	0.003364	0.0001869	0.744	0.75230	
Repeatability	60	0.015067	0.0002511			
Total	89	0.331640				

Source	VarComp
Total gage R&R	4.08E-04
Repeatability	2.36E-04
Reproducibility	1.71E-04
Operators	1.71E-04
Part-to-part	3.71E-03
Total variation	4.12E-03

Source	Std. Dev. (SD)	Study var. (5.15×SD)	% Study var. (%SV)	% Tolerance (SV/Tolerance)	
Total gage R&R	2.02E-02	0.103965	31.47	25.99	Look here first!
Repeatability	1.54E-02	0.079165	23.96	19.79	← Within operators
Reproducibility	1.31E-02	0.067391	20.40	16.85	← Between operators
Operators	1.31E-02	0.067391	20.40	16.85	
Part-to-part	6.09E-02	0.313594	94.92	78.40	
Total variation	6.42E-02	0.330379	100.00	82.59	
Number of distinct categories = 4 ← Must be > 5, Should be > 10					

2. *R-Chart by Operators:* The ranges should all be within the control limits and the R-bar should be relatively small. If any ranges are out of control, you have a repeatability problem. Find out what the operator is doing differently each time on the same part. In general, if there are less than five levels of range in the data, you may have a discrimination or rounding issue. Also, look for a difference in level between the operators—operators with ranges below the R-bar are more consistent than the others. Uncover their secrets!

3. *Xbar Chart by Operators:* More than half of the points should be outside the control limits, meaning that the R-bar is small as desired and hence the limits are tight. If more than half of the points are within the control limits, you have a repeatability problem. Next, look at the pattern of each operator's points. The patterns should be very similar. If they are not similar, you have a reproducibility problem. Here we must not confuse the control limits with those of a traditional process control. Here it is acceptable to be outside the limits, whereas in the traditional process we view them as unacceptable. In fact, if it occurs we say that we have assignable causes in the process.

4. *By Operators:* Depending upon your MINITAB configuration, you will get either a box plot or a dot plot for each operator and a red line will be drawn between the operator means. The red line should be nearly horizontal; if not, this could indicate a reproducibility issue.

5. *Operators × Part Interaction:* Different colored lines will represent the different operators and are drawn between the operator's mean for each part. The lines should be nearly parallel and close together. If not, this signals a potential interaction between the part and the operator. This means that there may be something about the part(s) that an operator had difficulty with. For example, a heavy part and a small operator may cause the results to be different than the other operators. Watch him or her measure that part again.

6. *By Parts:* The parts should be different from one another and the range of the parts should span the process width, Xbar ± 3 sigma. Choosing nearly identical parts can make the measurement system look bad, and choosing wildly different parts can make it look good. Therefore, the range of parts should be representative of the process output. You may have to sample the parts over days or weeks to include all normal sources of variation (operators, shifts, material lots, machines, temperatures, etc.). Watch the operator(s) measure

the affected part(s) again and look for the following possible causes for excessive bias and or linearity.

- For bias:
  - Instrument needs calibration
  - Worn instrument, equipment, or fixture
  - Worn or damaged master, error in master
  - Improper calibration or use of the setting master
  - Poor-quality instrument—design or conformance
  - Linearity error
  - Wrong gage for the application
  - Different measurement method—setup, loading, clamping, technique
  - Measuring the wrong characteristic
  - Distortion (gage or part)
  - Environment—temperature, humidity, vibration, cleanliness
  - Violation of an assumption, error in an applied constant
  - Application—part size, position, fatigue, observed error (readability, parallax)
- For linearity:
  - Instrument needs calibration, reduce the calibration interval
  - Worn instrument, equipment, or fixture
  - Poor maintenance—air, power, hydraulic, filters, corrosion, rust, cleanliness
  - Worn or damaged master(s), error in master(s)—minimum/maximum

The third area of error is in the area of human performance. The classic approach to human performance avoiding errors is in the area of mistake-proofing. Generally, mistake-proofing is a process improvement designed to prevent a specific defect from occurring. It comes to us from Japan and is known as *poka-yoke*. It means "fail safe." In fact, Shingo (1989) recognized three types of poka-yoke:

1. The *contact method* identifies defects by whether or not contact is established between the device and the product. Color detection and other product property techniques are considered extensions of this.
2. The *fixed-value method* determines whether a given number of movements have been made.

3. The *motion-step method* determines whether the prescribed steps or motions of the process have been followed.

Poka-yoke either gives warnings or can prevent, or control, the wrong action (Shingo 1989). It is suggested that the choice between these two should be made based on the behaviors in the process; occasional errors may warrant warnings, whereas frequent errors, or those impossible to correct, may warrant a control poka-yoke.

In the United States we have separated the term into two separate concepts. The first is *error-proofing*, associated with designs, and the second is *mistake-proofing*, associated with process. In either case, both focus on avoiding defects, errors, and nonconformances.

Error-proofing is based on two essential attitudes about human behavior:

1. Mistakes are inevitable.
2. Errors can be eliminated.

An error is any deviation from a specified manufacturing process and, as such:

- All defects are created by errors.
- Not all errors result in defects.

Example of an error: A hose clamp is not positioned correctly during the assembly. It is considered an error because it should have been designed in such a way that it could be fitted *only* one way, thereby eliminating the error. A defect, on the other hand, is the result of any deviation from product specifications that may lead to customer dissatisfaction. In the case of a clamp, the defect may be a leak. Remember that in order for something to be considered as a defect, (a) the product must have deviated from specifications, and (b) the product does not meet customer expectations. An obvious example of a defect is a door that is damaged due to hitting an object.

Mistake-proofing, on the other hand, maybe defined as the (a) application of tools and devices applied to a process to reduce the possibility of errors occurring, (b) application of tools and devices to a process to reduce the possibility of defects that have occurred continuing to the customer, and (c) use of functional design features to reduce the possibility of parts being assembled incorrectly.

Human performance may be evaluated from many sources, including assessment of the following documents:

- *Audit information:* Actions from audit information may be to (a) recognize and evaluate recurring nonconformances and (b) identify actions that could eliminate those nonconformances.
- *Advanced Product Quality Planning (APQP) guideline and forms:* Actions from APQP review may be to (a) develop a structured method for defining and executing the actions necessary to ensure a product satisfies the customer, and (b) develop an error-proofing methodology by identifying actions and defining methods resolving potential concerns.
- *FMEAs:* Actions from an FMEA may be to (a) recognize and evaluate the potential failure of a product or process and its effects, (b) identify actions that could eliminate or reduce the chance of the potential failure occurring, and (c) document the process.
- *8Ds:* Actions from 8D may be to provide a common process that effectively defines and resolves concerns and prevents their recurrence. It also attempts to
  - Increase management understanding.
  - Improve concern resolution and prevention.
  - Improve performance to quality/cost/timing.
  - Promote frank and open problem solving.
  - Provide automated computer support.
- *Best practices from similar components:* Actions from best practices maybe to (a) review and implement the best practices across the organization (in other words, institutionalize mistake-proofing processes) and (b) reevaluate practices that fall short of quality expectations.

The result of this assessment and other documents will result in better

- Design for manufacturing (DFM) strategies
- Design for assembly (DFA) strategies
- Error-proofing checklists
- Generic product design error-proofing strategies
- Lessons learned documents

In both error- and mistake-proofing there are a variety of ways to warn, avoid, control, and check for defects. Some of the most common devices are as follows:

- Sticky notes on components
- Electronic eyes
- Components that will only attach one way
- Limit switches
- Photo eyes
- Probes
- Proximity switches
- Profile plates
- Alignment tabs
- Broken tooling indicator
- Color striping

However, no matter what is used, the effect of the error- or mistake-proofing must be tracked and evaluated. A typical tracking form for evaluating error-proofing is the following:

- Identify problem.
- List possible errors.
- Determine most likely error (verify that this is the error.)
- Purpose multiple solutions.
- Evaluate solutions' effectiveness, cost, and complexity.
- Determine best solution. Verify that this solution will resolve the problem.
- Develop implementation plan.
- Analyze preliminary benefits.
- Develop plan for long-term measure of benefits.

On the other hand, a typical tracking and displaying the errors may be one or a combination of the following:

- Percentage of errors proofed
- Percentage of errors detected vs. percentage of errors prevented
- Percentage of errors attributed to the design
- Percentage of errors attributed to the process
- Usage of SPC; trend charts; Pareto diagrams; histograms, and so on

The process of actually performing a poka-yoke is a systematic approach and follows 12 steps:

1. Identify and describe the defect.
2. Show the defect rate by charting the defect occurrence over time.
3. Identify where the defect was found.
4. Identify where the defect was made.
5. Describe the current process where the defect was made by detailing the standard procedures used in the operation.
6. Identify any errors or deviations from process standards where the defect was made.
7. Use the five-why problem-solving technique to identify the root cause of the defect or error.
8. Develop ideas for improving the process using five-why tools to eliminate or detect the error.
9. Improve the process by creating an error-proofing device.
10. Measure/document results of error-proofing.
11. Standardize the improvement.
12. Ask where else this improvement could be used.

Though poka-yoke is an excellent approach to prevent defects in both design and process, there are other methodologies that may be used for specific applications. Some advanced human factors concepts are discussed next.

The Stereotyped Ordered Regression Model of Human Performance

The stereotyped order regression (SOR) model is based on the work of Bender (2004, 2006), which is based on SWOT (strengths, weaknesses, opportunities, and threats). SOR examines internal strengths (S) and external opportunities (O), and it combines internal weaknesses (W) with external threats (T) to formulate roadblocks (R). SOR = strengths, opportunities, and threats. It has been used with success in the leadership realm.

Rasmussen's Skill–Rule–Knowledge Model

The skills, rules, knowledge (SRK) framework or SRK taxonomy defines three types of behavioral or psychological processes present in operator

TABLE 3.3

Relationship Between the SRK Model and the Three Levels of
Control of Human Behavior

Knowledge	perception–recognition–decision–planning–execution
	perception–recognition–decision–execution
	perception–recognition–planning[a]–execution
Rules	perception–recognition–planning[b]–execution
	perception–recognition–execution
Skills	perception–execution

[a] The planning process adapts old cases to the new situation, and the adaptation is significant.

[b] The planning process adapts old cases to the new situation, and the adaptation is generally minor.

information processing (Vicente 1999). The SRK framework was developed by Rasmussen (1983) to help designers combine information requirements for a system and aspects of human cognition (the study of how the human brain thinks). In ecological interface design (EID), the SRK framework is used to determine how information should be displayed to take advantage of human perception and psychomotor abilities (Vicente 1999). By supporting skill- and rule-based behaviors in familiar tasks, more cognitive resources may be devoted to knowledge-based behaviors, which are important for managing unanticipated events. The three categories essentially describe the possible ways in which information, for example, from a human–machine interface, is extracted and understood. EID attempts to provide the operators with the necessary tools and information to become active problem solvers as opposed to passive monitors, particularly during the development of unforeseen events. Interfaces designed following the EID framework aim to lessen mental workload when dealing with unfamiliar and unanticipated events, which are attributed to increased psychological pressure (Vicente 1999). In doing so, cognitive resources may be freed up to support efficient problem solving. The progression of the SRK to execution is shown in Table 3.3.

Chris Wickens' Concepts on Mental Models

We describe the proximity compatibility principle (PCP) as one guideline to use in determining where a display should be located, given its relatedness to other displays. The PCP depends critically on two dimensions of proximity or similarity: perceptual proximity and processing proximity. *Perceptual proximity* (display proximity) defines how close together two display channels conveying task-related information lie in the user's multidimensional perceptual space (i.e., how similar they are). Thus, two sources will be perceptually more similar (in closer proximity) if they are close together, share the same color, use the same physical dimensions (e.g., both use orientation or length), or use the same code (e.g., both are digital or both are analog). For the designer, perceptual proximity is influenced by variation in where and how information sources are displayed, so this may be also referred to as display proximity.

Mental or *processing proximity* defines the extent to which the two or more sources are used as part of the same task. If these sources must be integrated, they have close processing proximity. If they should be processed independently, their processing proximity is low. The principle proposes a compatibility between these two dimensions. If there is close processing proximity, then close perceptual proximity is advised; conversely, if independent processing is required, distant perceptual proximity is prescribed (Carswell and Wickens 1995).

James Reason's Swiss Cheese Model of Error

The Swiss cheese model of accident causation is a model used in the risk analysis and risk management of human systems. It likens human systems to multiple slices of Swiss cheese, stacked together, side by side. It was originally propounded by British psychologist James T. Reason (1990) and has since gained widespread acceptance and use in health care, in the aviation safety industry, and in emergency service organizations. It is sometimes called the *cumulative act effect.*

Reason (1995, 1997, 2000) hypothesized that most accidents can be traced to one or more of four levels of failure: organizational influences, unsafe supervision, preconditions for unsafe acts, and the unsafe acts themselves. In the Swiss cheese model, an organization's defenses against failure are modeled as a series of barriers, represented as slices of Swiss cheese. The holes in the cheese slices represent individual weaknesses in individual parts of the system and are continually varying in size and position in all slices. Reason's own words as reported in (Smith, Frazier, Reithmaier and Miller 2001, p.10) claims that the system as a whole produces failures when all of the holes in each of the slices momentarily align, permitting "a trajectory of accident opportunity," so that a hazard passes through all of the holes in all of the defenses, leading to a failure (Smith et al. 2001; Wilson et al. 2002; Amos and Snowden 2005).

SUMMARY

In this chapter we discussed the execution of a determined decision by using PM, recognizing risk, and being cognizant of measurement error. In the next chapter we will focus on the opportunities for innovation.

REFERENCES

Amos, T. and P. Snowden. (2005). "Risk management." In J. B. Adrian and T. James.

Amos, T and P. Snowden (2005). "Risk management." In Adrian J. B. James, Tim Kendall, and Adrian Worrall *Clinical Governance in Mental Health and Learning Disability Services: A Practical Guide*. London: Gaskell.

Alderson, M. (June 14, 2008). Risk assessment. *Financial Times*. http://www.ft.com/cms/s/2/b37b1722-38da-11dd-8aed-0000779fd2ac.html#ixzz1S1c8CBgQ

Assessment of risk – others have also been identified: http://www.hse.gov.uk/pubns/indg163.pdf

Bender, S. L. (2004). *SOR analysis: Beating SWOT analysis in performance management*. http://www.sharonbender.com. Accessed on November 20, 2010.

Bender, S. L. (2006). *SOR analysis: An opportunities-based intervention*. http://www.sharonbender.com. Accessed on November 20, 2010.

Brown, S. P. Implement Enterprize Risk Management. http://ezinearticles.com/?Implement-Enterprise-Risk-Management&id=1140844

Carothers, D. http://hubpages.com/hub/From-Checklists-to-Experts-The-Risk-Identificaton-Phase

Carswell, C. M. and C. D. Wickens. (1995). "The proximity compatibility principle: Its psychological foundation and relevance to display design." *Human Factors* 37, 473–494.

Common risk: http://www.vectorstudy.com/management_topics/risk_management.htm

Courtney, R. (1970). http://iwenger.com/risk_management_overview

Crockford, Neil (1986). *An Introduction to Risk Management* (2 ed.). Cambridge, UK: Woodhead-Faulkner. p. 18.

Cve: http://cve.mitre.org

DaimlerChrysler Corporation, Ford Motor Corporation, and General Motor Corporation. (2002). *Measurement system analysis (MSA)*, 3rd ed. Southfield, MI: Automotive Industry Action Group.

Darwin's theory. http://www.darwins-theory-of-evolution.com/.

Dorfman, M. S. (1997). *Introduction to risk management and insurance*, 6th ed. Upper Saddle River, NJ: Prentice Hall.

Franck, C. (2008). Business risk management. http://www.aiu.edu/publications/student/english/Business%20Risk%20Management.html

Freestone, R., Thompson, S., & Williams, P. (2006). "Student experiences of work-based learning in planning education." Journal *of Planning Education and Research, 26*(2), 237–249.

Gaedeke, R, M. and D. H. Tootelian. (May 22, 1989). "Gap found between employers' and students' perceptions of most desirable job attributes." *Marketing News*, p. 42

Hajar, H. http://www.aiu.edu/applications/DocumentLibraryManager/upload/Houssam%20Hajar.pdf

Hajar, H. (b). Financial Risk Management. http://www.aiu.edu/publications/student/english/Financial-Risk-Management.html

Halon. http://www.nationmaster.com/encyclopedia/HALON

Hubbard, D. (2009). *The Failure of Risk Management: Why It's Broken and How to Fix It.* New York: John Wiley & Sons.

Institute and the American Management Association. (2006). The ethical enterprise. http://www.amanet.org/HREthicsSurvey06.pdf

Kelvin. http://zapatopi.net/kelvin/quotes/

Lea, W. (2007). *Managing Successful Projects with PRINCE2.* London: The Stationary Office (TSO).

Lippitt, M. (August 2007). "Fix the disconnect between." *T & D, 61*(8), 43–45. See also: http://news-business.vlex.com/vid/fix-the-disconnect-between-64106197

Mobley, R. K. What is Risk Management? http://www.reliabilityweb.com/index.php/print/what_is_risk_management

NASA. (2008). http://esciencenews.com/sources/la.times.science/2008/08/23/off.course.rocket.destroyed.nasa

Nutt, P. C. (2002). *Why Decisions Fail.* San Francisco: Berrett-Koehler Publishers.

Prince 2: http://www.jiscinfonet.ac.uk/infokits/project-management/prince2.pdf

Prince 2 (a) http://www.prince2.com/us/

Prince 2 (b) http://www.prince2.com/prince2-process-model.asp

Prince 2 (c) http://www.nikfi.com/WhatisPRINCE2-ThePRINCE2ProcessModel.htm

Prince 2 (d) http://eidzina.blogspot.com/2010/09/prince2-open-your-mind.html

Prince 2 (e) http://www.prince2training.com/index.php?module=prince2processes

Rasmussen, J. (1983). "Skills, rules, and knowledge: Signals, signs, and symbols, and other distinctions in human performance models." *IEEE Transactions on Systems, Man, and Cybernetics* 13(3): 257–266.

Reason, J. (1990). *Human error.* Boston: Cambridge University Press.

Reason, J. (1995). "A system approach to organizational error." *Ergonomics* 38: 1708–1721.

Reason, J. (1997). *Managing the risks of organizational accidents.* Ashgate, NY: Aldershot, UK.

Reason, J. (2000). "Human error: Models and management." *British Medical Journal 320* (March), 768–770.

Risk Assessment. http://en.wikipedia.org/wiki/Risk_assessment

Risk Management. http://en.wikipedia.org/wiki/Risk_management

Risk Management Process (a). http://www.method123.com/risk-management.php

Risk Management Process (b). http://www.buzzle.com/articles/risk-management-process.html

Risk Management (c) http://www.wildwhisper.com/risk_management.html

Shingo, S. (1989). *A study of the Toyota production system.* New York: Productivity Press.

Smith, D. R., D. Frazier, L. W. Reithmaier, and J. C. Miller. (2001). *Controlling pilot error.* New York: McGraw-Hill Professional.

Smith, D. R., Frazier, D., Reithmaier, L. W. and Miller, J. C. (2001). *Controlling Pilot Error.* McGraw-Hill Professional. p. 10

Stamatis, D. H. (1997). *TQM engineering handbook.* New York: Marcel Dekker.

Stamatis, D. H. (2003a). *Failure mode and effect analysis: FMEA from theory to execution,* 2nd ed. Rev and expanded. Milwaukee, WI: Quality Press.

Stamatis, D. H. (2003b). *Six Sigma and beyond: Design of experiments*. Vol. 5. Boca Raton, FL: St. Lucie Press.

Stamatis, D. H. (2003c). *Six Sigma and beyond: Statistics and probability*. Vol. 3. Boca Raton, FL: St. Lucie Press.

Standards Australia and Standards New Zealand (2004)/ AS/ANZ 4360:2004 Risk Management. Sydney: NSW.

Standards Australia and Standards New Zealand (2005) (Reissued incorporating Amendment No. 1: December 2005). Risk Management Guidelines Companion to AS/NZS 4360:2004. Sydney. NSW. (Originated as HB 142—1999 and HB 143:1999. Jointly revised and redesignated as HB 436:2004).

Standards Australia/Standards New Zealand (2004a). Originated as HB 221:2003. Second edition 2004. Jointly published by Standards Australia International Ltd. Sydney. NSW.

Support. http://www.h3rcleanagents.com/support_faq_2.htm

Thornton, J. (2000). *Pandora's poison*. Boston: MIT Press.

Vicente, K. J. (1999). *Cognitive work analysis: Towards safe, productive, and healthy computer-based work*. Hillsdale, NJ: Lawrence Erlbaum Associates.

Wilson, J. H., A. Symon, J. Williams, and J. Tingle. (2002). *Clinical risk management in midwifery: The right to a perfect baby?* Philadelphia: Elsevier Health Sciences.

SELECTED BIBLIOGRAPHY

Bradley, K. (1997). *Understanding PRINCE 2*. Poole, Dorset, England: SPOCE Project Management Limited.

Baird, D. C. (1988). *Experimentation, an introduction to measurement theory and experiment design*, 2nd ed. Upper Saddle River, NJ: Prentice-Hall.

Bayley, C. (2004). "What medical errors can tell us about management mistakes." In *Management mistakes in healthcare: Identification, correction, and prevention*, edited by P. B. Hofmann and F. Perry. Boston: Cambridge University Press.

Bork, P. V., H. Grote, D. Notz, and M. Regler. (1993). *Data analysis techniques in high energy physics experiments*. Boston: Cambridge University Press.

Borodzicz, Edward (2005). *Risk, Crisis, and Security Management*. New York: Wiley.

Daume, H. (2002). "Frictionless project management." *Quirk's Marketing Research Review* January, 44–47.

Gorrod, M. (2003). *Risk management systems: Technology trends: Finance and capital markets*. New York: Palgrave Macmillan.

Haughey, D. Project Planning a Step by Step Guide. http://www.projectsmart.co.uk/project-planning-step-by-step.html

Hinckley, C. M. and P. Barkan. (1995). The role of variation, mistakes, and complexity in producing nonconformities. *Journal of Quality Technology* 27(3): 242–249.

http://bizwizards.net/projectfeasibility.html

http://en.wikipedia.org/wiki/Risk_management

http://issuu.com/siva_asokan/docs/management

http://vsa2008wikiworkshop.pbwiki.com/Plan-the-project

http://www.ghrogroup.com/

http://www.jiscinfonet.ac.uk/InfoKits/risk-management

http://www.projectsmart.co.uk/project-planning-step-by-step.html
http://www.theirm.org/aboutheirm/ABwhatisrm.htm
http://www.vgic.com/Default.aspx?tabid=132
Hubbard, Douglas (2009). The Failure of Risk Management: Why It's Broken and How to Fix It. John Wiley & Sons.
ISO/DIS 31000 (2009). Risk management — Principles and guidelines on implementation. International Organization for Standardization. http://www.iso.org/iso/iso_catalogue/catalogue_tc/catalogue_detail.htm?csnumber=43170.
ISO/IEC Guide 73:2009 (2009). Risk management — Vocabulary. International Organization for Standardization. http://www.iso.org/iso/iso_catalogue/catalogue_ics/catalogue_detail_ics.htm?csnumber=44651.
London, C. (2002). "Strategic planning." *Quality Progress* August, 26–33.
Marshall, L. Time Management in a Multi-Project Environment: Time and task management in a multiple concurrent project environment. http://www.projectsmart.co.uk/project-planning-step-by-step.html
Meiners, H. F., W. Eppenstein, and K. H. Moore. (1969). *Laboratory physics.* New York: John Wiley & Sons.
Nikkan Kogyo Shimbun, Ltd. (1988). *Poka-yoke: Improving product quality by preventing defects.* Portland, OR: Productivity Press.
Roberts, T. Project Plans: 10 Essential Elements. http://www.projectsmart.co.uk/project-plans-10-essential-elements.html
Shingo, S. (1985). *Zero quality control: Source inspection and the poka-yoke system,* translated by A. P. Dillion. Portland, OR: Productivity Press.
Stulz, R. M. (2003). *Risk management and derivatives,* 1st ed. Mason, OH: Thomson South-Western.
Swartz, C. E. (1973). *Used math, for the first two years of college science.* Upper Saddle River, NJ: Prentice-Hall.
Swartz, C. E. and T. Miner. (1977). *Teaching introductory physics: A sourcebook.* Melville, NY: American Institute of Physics.
Symonds, M. How To Regain Control Of Your Project. http://www.projectsmart.co.uk/project-planning-step-by-step.html
Symonds, M. Project planning essentials. http://www.projectsmart.co.uk/project-planning-step-by-step.html
Taylor, J. R. (1962). *An introduction to error analysis.* Sausalito, CA: University Science Books.
Taylor, J. R. (1982). *An introduction to error analysis: The study of uncertainties in physical measurements.* Herndon, VA. University Science Books.
Thomsett, R. (2002). *Radical project management.* Upper Saddle River, NJ: Prentice Hall PTR.
United States Department of Labor Occupational Safety and Health Administration. (April 2004). *EPA general risk management program guidance.* Boulder, CO: U.S. Printing Office.
Vicente, K. J. and J. Rasmussen. (1990). The ecology of human–machine systems II. Mediating "direct perception" in complex work domains. EPRL-90-91. University of Illinois at Urbana–Champaign.
Vicente, K. J. and J. Rasmussen. (1992). "Ecological interface design: Theoretical foundations." *IEEE Transactions on Systems, Man and Cybernetics* 22, 589–606.
Young, H. D. (1962). *Statistical treatment of experimental data.* New York: McGraw-Hill.

4

Innovation

In the last chapter we addressed project management, risk, and measurement error as the key characteristics to execute a decision. In this chapter we will focus on innovation and how an organization uses innovation for improvement.

Innovation is about new revolutionary ideas and their execution. Therefore, the best way to bring fresh eyes to any problem is to bring in new kinds of people. When it comes to innovation, no one is too different or revolutionary for any level of the organization.

This, of course, for most organizations is an oxymoron. Organizations preach that they want an "out of the box" mentality and push the idea of "being different," but when those ideas come forth, they cannot depart from their deeply ingrained beliefs and practices about how to treat people, make decisions, and structure work. They always defend the culture with "it has never been done here before like this."

Webster's Dictionary defines *innovation* as "a new idea or method or device." That definition is rather vague; it implies that anything new is innovative. That is simply not the case. Painting a blue widget red or making it round instead of square may have merit, but it is not necessarily innovative. Innovation comes from the Latin word *novare*, which includes, as part of its meaning, "to change, to alter, to invent." Innovation carries with it a newness that comes from a fresh approach. It is a way to think out of the box.

In a global economy, the cost of production is often a driving factor in deciding where a product is made. Innovation directly influences the amount of economic power a region wields. A case in point is the power of China. Its cheap labor gives it a leadership role when it comes to exports. At one time, cheap labor, and the power that went with it, was found in South

Korea, Japan, and Mexico. The competitive advantage that the United States has is its ability to innovate. That advantage finally may be challenged.

Gupta (2007) has suggested that this challenge is real, because innovation is a process that can be understood and, hence, taught and refined to improve the results. In understanding and implementing such a process, an individual, company, or country can maintain a leadership position in business, education, manufacturing, or whatever their chosen field. Therein lays the challenge for U.S. industries, especially manufacturing.

If innovation can be learned here, it can be learned in other countries. In regions such as China, Vietnam, Cambodia, India, Eastern Europe, and elsewhere where there is a high level of education, the United States could find itself in a tight race in which it has long held the lead.

To survive, companies must review everything they do and how they do it. They must be responsive to both customer requirements and increasing competitive pressure. They must be able to innovate more (quantity) than the competition as well as more often than the competitors and satisfy the customers no matter where they are. This means that organizations in the age of the Internet must demand a revolution in organizational structure and policies. They must demand a talent-based corporate approach and strong leadership as a driver of change. All this will maximize innovation through people and facilitate the development of a framework that encourages independent thinking and entrepreneurial action within the organization.

DESIGN

Innovations in the design stage generally mean moving ideas from marketing to engineering, developing new products and producing them quickly to satisfy specific markets. The faster the process is, the better it is. This sounds very easy; however, in the real world it is very difficult to accomplish because there is no silver bullet for success in innovation.

In general terms, innovation has three models:

1. *Need seekers*: Companies following this strategy actively engage current and potential customers to shape new products, services, and processes so that they can be the first to market with those products.
2. *Market readers*: Organizations in this category carefully watch their markets but mainly focus on creating value through incremental change.

3. *Technology drivers*: Organizations in this category, not unexpectedly, use their inherent technical capabilities to leverage R&D (research and development) spending to drive breakthrough innovation and incremental change to solve the unarticulated needs of their customers.

For any innovation strategy to work, regardless of which model one follows, it must be aligned with overall corporate objectives. Experience has shown that organizations that follow that simple rule showed 400 percent higher operating income growth and 100 percent higher shareholder returns than less aligned companies. The reason for this astonishing growth is that this tight alignment engages the customer base of the organization directly.

By clarifying corporate objectives, the customer's spoken and unspoken requirements are heard. This more or less gives designs a better chance for success. It is the efficiency of the functional interfaces between engineering, product development, sales and marketing, etc., that determines a company's innovation performance, including when to stop a program from going forward.

This, of course, takes discipline to constantly evaluate the value proposition of the product, the parameters of the business case, and any changes that occur during the development process. The process of adjustment will tell you when to proceed and when—despite the amount of money spent— to kill a product. Often this is more difficult than it appears, because sometimes decisions are made on political grounds and power plays regardless of what the data say. In fact, the bias to continue development of a product despite the fact that it is off-target is quite overpowering.

In terms of spending for innovation and R&D, the auto industry is the only sector that continues to decline. Spending is dropping partly due to the maturity of the auto industry in all aspects of its cost structure. Efficiency improvements in terms of the product development process—emphasized by the move to fewer and more flexible platforms—and the use of low-cost sources for engineering and technical services at the Tier 1 level, coupled with a product offensive pushing more models into more finely defined segments, have kept the spending low. By comparison, Toyota's fast, efficient product development system lets it make product investments that other companies find economically unattractive, and therefore receive more benefits from using their own product development system known as the "Toyota Production System." It is the nature of how they approach the process of design choices and resolutions, when to make tradeoffs and what tradeoffs to make, and how this is a natural part of the process instead of a calendar-driven, stage-gate process, that explains much of their success. It also may be proof that, as with the high-leverage innovators, spending less to get more is the better choice.

We do not suggest that copying Toyota's system is the answer. To the contrary, successful companies do not copy. Rather, they emphasize how to capture the value of innovation in the most compressed time available. This implies what kind of priorities the organization wants to pursue, what is achievable, and how much of the value you can capture given the cycle time compression and the number of fast-following new entrants there are in the market. To make the point, let us see Toyota, which has an 18- to 24-month product development cycle, compared with over 40 months for the typical American company. Toyota is very good at not being first while capturing a lot of the segment value by being quick to market with a product that capitalizes on its reputation for quality, durability, and reliability. The focus is on reusability and carryover. American companies want to be the first in the market and claim technological innovation at the expense of unreliable products with quality issues.

Obviously, to effectively compete, each organization must create an economically advantaged system based on its own strengths and weaknesses. This implies evaluating how to change the process currently used, determining which fundamentals have to be changed, and studying what capabilities are currently missing and how they can be brought into the system. Organizations must understand that you cannot borrow innovation. Rather, to be successful, you have to determine what works best for you. A typical comparison and evaluation of an organization planning for innovation is shown in Table 4.1.

MANUFACTURABILITY/ASSEMBLY

Globalization, time-to-market, and cost pressures all create challenges in general assembly manufacturing. Companies must manage their planning process for greater efficiency and consistency across globally dispersed plants, balancing production activity and resources even as product customization increases. Though each company faces a different set of challenges, at the highest level there are four areas of concern that are common to most companies:

TABLE 4.1
Innovation Strategy and Critical Dimensions

Innovation Strategy	Ideation	Project Selection	Product Development	Commercialization	Critical Dimensions
Need Seekers					
Identify unmet customer needs through direct feedback and strive to be the first to market with breakthrough products Example: DeWalt (power tools)	Gather customer insights and analyze customer needs Segment customer base	Rigorously manage return on innovation investment	Design products that respond to customers' priorities	Successfully launch, position, and price wholly new products	First to market Direct customer insight Technology forward
Market Readers					
Focus on incremental changes to products and use a second-mover strategy to keep risk low Example: Plantronics (audio equipment)	Conduct market research Gather competitive intelligence	Maintain strong process discipline	Bring products quickly to market with an emphasis on increased modularity and simplicity	Carefully manage product life cycle and retirement	Incremental change Fast follower Indirect customer insight Market back
Technology Drivers					
Rely on technological breakthroughs from internal R&D efforts and seek to meet their customers' unarticulated needs Example: Siemens (engineering and electronics)	Scout new technologies Map emerging technologies and analyze trends	Manage risks	Test rigorously for quality	Capture customer feedback	Breakthrough innovation First to market Technology forward Indirect customer insight

- Poor process and workflow coordination
- Planning and production resource constraints
- Insufficient visibility into variable plant capabilities
- Regulatory requirements

In manufacturing today, innovation is not just a way of life, it is a means of survival. Key to that survival is partnering with strong suppliers who will go the distance to meet critical production requirements.

The future will bring new products, tougher quality standards, and fierce competition and cost concerns. The future will mandate more creative solutions to assembly and test. Whether synchronous or nonsynchronous assembly is required, the need for integration in speed, accuracy, and reliability will always be in the forefront of innovation.

This integration will be in conjunction with (a) design for assembly (DFA), which is a process that minimizes product cost by analyzing assembly time, part cost, and the assembly process at the product's design stage; and (b) design for manufacturability (DFM), another assembly tool, which addresses material and tooling in product development.

Manufacturers that combine DFA with DFM (DFMA) will significantly improve product quality and time to market.

SERVICE

People think that customer service will fall in place, but customer service planning is as important as profit margin because it affects profit margin. One must know how to respond—deliveries will go wrong, a customer will feel offended. Therefore, having a customer service plan and training in place is vital.

To talk about *customer service* or *customer satisfaction* is abstract. However, if we are interested in innovation of service, we must be aggressively involved with the customer to find what the "thing" is that will make him buy and gain his loyalty.

Customer service guidelines can be quite elaborate, but they do not necessarily need to be. The bottom line is to give customers an easy, clear way to provide feedback. You could, for example, include a service reply card with every receipt or invoice, mandate that phone calls are answered within three rings and track the number of calls rolling into voicemail, run customer surveys, or simply ask customers what they think. On the other hand, it is imperative to understand that small-business people do not need fancy business programs. They just need to stick out their hand and say, "Hi, I'm the owner. How am I doing?"

So, what do customers want? They want a terrific product or service, certainly. But they want a great deal more. The right merchandise or transaction is just "the basic ticket into the game." Customers also want friendliness, understanding, and empathy with their situation. They want fair treatment as well, some feeling of control in the transaction outcome, or to be told their options, if that is necessary. And they want still more!

As an owner of a small retail computer store, a consulting business, a founder of a nonprofit organization, and a professional quality consultant for over 30 years, I have come to the conclusion that what customers really want may be summarized in the following five items:

1. Caring employees. That means that customers want employees who care enough to say "This is not right for you" or "I have something that fits your needs better" even if the price is less. They put the customer ahead of the sale.
2. Good value and good product knowledge.
3. Knowledgeable and available sales staff. When a customer wants to see a salesperson, he wants to see one right now, yet he or she hates when they hover. This may sound like an oxymoron, but that is precisely why we need to have a balancing strategy with the intent to satisfy.
4. Well-organized and easy-to-find merchandise.
5. Fast finish. Customers repeatedly have communicated that when they want to check out, they want to check out now. Customers hate not having enough cashiers to make checkout quick. They also do not appreciate when the employee at a customer service counter, for example, does not know how to exchange a return item, and so on.

So, the fundamental question is, how do you know what each particular customer wants most? The answer, quite simply, is start by listening and putting the customer first. That may be done with a strategy to

- Stand above the crowd (competition) and take responsibility for what the customer really wants and give it to them. Do not stand for the basic service basics. Get to know your customers and cultivate their loyalty by maintaining contact with them. Offer legitimate guarantees and offer rewards. Treat them like royalty.

- Empower employees to handle customer complaints and situations appropriately. This is especially true for customer service representatives (CSRs). It is very frustrating when a CSR cannot help you on the spot—even for items less than $50.00. Instead, they have to escalate the problem to higher levels of CSR management. It is waste of time for the organization and very frustrating for the customer to have to repeat the problem at hand several times over.
- Regain angry customers by providing answers to their concerns as well as appropriate and applicable avenues for them to participate in your business. Make amends.

Whereas generally people have a good idea of what technological innovation is, most people have never thought about service innovation, because it is more hidden. Service innovation is a knowledge management strategy for service and support organizations. Service innovation is not just a matter of "Aha!" ideas. It is rather a process that requires a disciplined approach to rigorously identify and execute the most promising ideas. It defines a set of principles and practices that enable organizations to improve service levels to customers, gain operational efficiencies, and increase the organization's value to their stockholders.

The latest data from the Bureau of Labor Statistics (December 2005) project that the service-providing sector of the U.S. economy will see the highest employment growth by 2014. In fact, the statistics show that the ten industries with the most dramatic salary growth are all in this service-providing sector, ranging from employment services to education to health care. In contrast, employment in the manufacturing sector is expected to drop 5 percent by 2014, and goods-producing industries will decline to 13 percent of total American employment in the decade 2004–2014, down from 15 percent in 1994–2004. And the trend is global: service workers now outnumber farmers for the first time, according to the January 2007 issue of *Global Employment Trends*, a newsletter published by the International Labor Office.

The basis for any innovation success in overcoming customer's or client's objections is the degree to which you are considered a valued, trusted partner by the customer/client. Partnering involves a long-term business relationship built on the value you bring to the table and your commitment to the client's success. It is essential that you add value in every customer/

client meeting, telephone call, and email and deliver excellence in service to establish credibility and trust with your client (Jones & Samalionis 2008).

Because many organizations must operate with limited resources and competing demands, customers/clients are often in a rush to take action to solve problems. The painful consequences for you and the customer/client are often that the same business issue must be attacked multiple times. Solving the problem the first time is a valuable service that performance analysis can provide.

Your analysis strategies should never deviate from your ultimate goal—to provide reliable, quality data that help your customer/client make sound business decisions and that support improved business results. Key elements toward this goal include the following:

- Remain a true strategic partner before, during, and after data analysis. Make sure that you do not take over the ownership of the business problem from the customer/client relationship. Your role as a strategic business partner is to provide reliable data to help your customer/client make sound business decisions needed to achieve business goals.
- Know and build upon your strengths and develop new competencies. Most of the methodology and process knowledge competencies can be obtained through education and on-the-job application. You must demonstrate strategic thinking, visionary thinking, and the ability to systemically "peel the onion" to clearly identify the real problem.
- Work as part of a team with performance partners in human resources, including organizational development, and with Six Sigma experts. Be sure to include a star performer and subject matter expert from the customer's/client's business unit to help you interpret the data and gain their buy-in and trust.
- Gather as much internal and external information as you can on the performance gaps. Prepare a gaps table for the customer/client to show the target of addressing the gaps and achieving desired goals.
- Ask thorough questions that are pretested to obtain objective and reliable data. Be sure to ask enough open-ended and probing questions to uncover the root causes and potential solutions for each cause.
- Stay true to your findings based on the data. Precision in data collection and analysis will lead to the appropriate recommendations. Do not identify a cause or solution unless it is supported by your data. Avoid presenting your personal opinions and preferences.

- Have confidence in the methodology and your findings. This will be apparent in your meetings and presentations with the customers/clients.
- Use a concise, easy-to-follow format for your findings report. Prepare an executive summary that includes a review of the methodology, key data findings, and recommendations. Include wording from the customer's/client's agreement if possible. Prepare a binder with detailed data that will help you answer all pertinent questions.
- Do not provide your findings report to your customer/client in advance. Seize the opportunity to present your information in person. The passion and additional information you will add in this face-to-face meeting will increase understanding and buy-in. It also decreases the chances of your data being misinterpreted and implemented incorrectly.
- Be intentional about how you manage customer/client expectations prior to presenting your data report. Start the presentation by reminding the client about your analytical rigor and objectivity to eliminate objections to the data during the meeting and to increase client support.

Now that we have addressed some of the issues of customer service and satisfaction, the strategy for innovation, and the key elements of that strategy, let us suggest a model suggested by Jones and Samalionis (2008) and expanded by the author. This simple, basic model for a service innovation includes the following steps:

1. *Develop insight about the market:* Develop insights about customers, the business, and technology in parallel. Simply observing customers may not be enough to drive the kind of innovation that changes markets. Inspiration can come from many areas, so be as insightful as you can about alternative business models, market landscapes, and operational and technology infrastructure. Innovations will come from the union of these perspectives and will be successful when aligned with customer needs.

   Develop frameworks that clearly describe the "pain point" and the opportunity space. What you do with market insights is more important than the insights themselves; too often, companies do not take advantage of their full potential. Excitement over customer insights alone can lead teams to jump the gun and start brainstorming service solutions that solve only specific issues. This tends to result in incremental service improvements rather than the more substantial leaps the team is looking for.

It takes time for a team to immerse itself in the nuances of a problem. But it is critical to take that time to develop meaningful frameworks that can structure ideation. A team knows that it is ready to move on to ideation and prototyping when it sees an opportunity for a radically different way to serve customer needs.

2. *Create radical value propositions:* Radical innovation is about acquiring new customers and tapping underserved markets, as well as retaining those people once they become customers. Giving people a reason to try your service in a crowded marketplace requires going a step above what they experience with their current service. And if what you are offering is a new class of services—think Zipcar,[1] for example—then you will have to help your customers recognize the value of trying something new. Sometimes, radical services fill an obvious gap in the marketplace—think Google,[2] 411, and Short Message Service (SMS) information services. At other times, they help steer markets in new directions by capitalizing on existing but fragmented behaviors— think Apple's iPod and iTunes (Jones & Samalionis 2008).

Prototype extreme service propositions early to stretch the organizational mind-set. Quick, low-cost mockups allow emerging ideas to be expressed, explored, modified, and shared with customers, experts, and stakeholders in a very tangible and emotive way. They encourage informed decision making more than a paper description could ever do, and they encourage the idea to continually evolve. Because they often deal with the intangible, service ideas may also require simulation or even the acting out of a scenario. For example, simulating a customer's experience of interacting with a service can be an invaluable tool.

Experiencing prototypes that look like and behave like, but are not built like, the innovative new service allows a diverse range of customers, as well as stakeholders (those involved with brand, marketing, technology, customer care, delivery, and so on), to engage with and build on the new service from their specific perspectives. A good prototype will prompt questions around consumer desirability, business viability, and technical feasibility. A classic example is Bose speakers. Until the introduction of Bose speakers, the general thought was that the bigger the speaker (actually the bigger the magnet), the better the quality of the sound. That notion of quality was shattered with the miniature size of the Bose speaker with a superb sound quality.

3. *Explore creative service models:* Innovations that have the ability to change the marketplace usually require radical or fundamental changes within an industry, as well as creative solutions to make these new service offerings viable from a business perspective and feasible from a technology perspective. Challenge the existing operational realities. Successful innovation requires business and technology team members to be as creative as their design counterparts—think of Google's service model, which allows it to monetize offerings through ad revenue without compromising the value provided by the service. It is critical to remember that the success of these offerings rejuvenated the advertising industry, which was not showing too much promise at the time. Facebook is still refining how it will use information about users' interests and activities to support highly targeted advertising, and it hopes to offer a similarly revolutionary ad scheme as Google did with AdWords.

   Champion customer desirability as the reality of viability and feasibility are considered. It is easy to revert to traditional benchmarks and models, but doing so dramatically reduces the innovation from radical to incremental. Championing the desirability of an innovation forces the organization to build new constructs that will nurture radical innovations. And this is not an easy task.

4. *Bend the rules of delivery:* Get permission to fail. Radical innovation is risky in the sense that getting it right the first time, every time, is highly unlikely. Companies should expect failures as a part of the innovation process. Companies that do not have failures go out of business in due time. Teams need to have buy-in from leaders so that they feel confident trying new service concepts that have many unresolved questions. Teams that are afraid to fail make radical innovation, by definition, impossible. Set up for successful experimentation by getting buy-in from leadership, and then use that buy-in to get permission to fail (Jones & Samalionis 2008).

   Design new metrics for measuring success. Always remember that things that are measured do get done. On the other hand, remember that rules about metrics can be a huge barrier to innovation. This is especially true within service organizations that have adopted Six Sigma and/or Lean Six Sigma methodologies. Funding guidelines that work well for the evolution of incremental improvements to services are often at odds with the scale and ambiguity of radical innovation. Radical service concepts may not have a business case that meets Six Sigma guidelines; waiving those criteria can open up opportunities that would normally be squelched. Innovation efforts are more likely to be successful if they are funded and measured separately from the rest of the organization.

5. *Iteratively pilot and refine the new service:* Radical innovation is inherently risky because it involves new-to-the-world offerings. Piloting a service is the best way to manage this risk—before it is scaled. But most service organizations are reluctant to expose their intent to the market. Therefore, they are reluctant to pilot in order to protect their first-mover advantage. That reluctance needs to be balanced against the advantages that pilots offer in informing investment decisions.

Do not wait for the service to be perfect; get comfortable with beta testing or a pilot study. Radical innovation is fundamentally based on evolving customer behaviors and market trends. These changes are hard to predict accurately, and the success of a service can hinge upon a small nuance that is hard to pinpoint unless it is highlighted in a pilot. A works-like prototype can be easily piloted on a small scale to drastically reduce development cost, and it allows for iterative refinement that is critical to risk management.

The market landscape for services is evolving constantly and rapidly. Both the market and customers expect nimbleness when it comes to innovative services.[3] Frequent and radical innovations are keys to being relevant in such a landscape (Jones & Samalionis 2008).

OUTSOURCING

Not too long ago, outsourcing was the talk of the big business, primarily for manufacturing. Outsourcing started as a way to optimize costs for manufacturing organizations within a particular country. As time went on, this practice began to expand into the international markets (globalization concept), such as Mexico, Brazil, Eastern Europe, and especially China.

In the beginning, transfer of production was strictly cost oriented; however, soon the organizations that part took in these endeavors found that there were more benefits than strictly savings on particular production parts or subassemblies or even assemblies. The benefit of opening new markets for their products was even bigger than the savings of labor and other costs associated with their outsourcing.

For example, in China alone, with a population of over 1.4 billion, we see drastic changes not only from manufacturing production ability but also consumption. Its economy has become the fourth largest, with an average gross domestic product (GDP) increase of 9 percent every year since the mid-1990s. Consumer spending is rising by about 14 percent per year, with the number of people who can buy luxury goods currently at about 200 million (equivalent to two-thirds of the U.S. population or the entire population of France, Spain, and Germany combined). In Beijing alone one can find Tiffany's jewelers and a Porsche showroom, as the country's new upper classes develop an appetite for Western luxuries. Tastes are changing and organizations that have the vision to take advantage of such changes will be rewarded. Companies like Microsoft, NEC, Ford Motor Company, General Motors, and many others are already taking advantage of the growth opportunities in China.

By 2012, most of the organizations will produce products and services in the Asia market to the tune of about 36%, and 51% of that will go to China. In the case of smart phones the projection is even more drastic as the forecast of growth is about 90% over 2010 (Administration, 2010). To be sure it is enticing to jump into the flurry of action and be part of this phenomenal growth. Innovation at its best!

One therefore must be very careful, because it is not all clear sailing. There are scams, pitfalls, and dangers just like anywhere else. First and foremost are the language and cultural barriers to negotiate, as well as the fear of corruption and blackmail. The enticement of low wages and inflation are also a concern. In the case of wages, it is true that in the coastal strip, especially around Shanghai, Guangzhou, and Beijing, wages are creeping up; however, they are still quite low. As for inflation, it is all relative. Yes, there is high inflation due to high wages, but it is from a very low base. Wages, for example, in one factory in Shandong are about US$200.00 per month; in Chongqing an engineer with a master's degree averages about US$550.00 per month—and that is working 10–12 hours a day and quite often on weekends.

As of late, however, we see outsourcing in the world of service providers—so much so that international consulting firms set up dedicated services that enable large organizations to have information technology, engineering, payroll, and other back-office functions managed by teams of specialists. This is occurring in the Philippines, India, China, South Korea, Brazil, and many others.

However, with the arrival of the Internet, more smaller businesses are taking the opportunity to gain from the concept. Increasing numbers of growing businesses are, for example, avoiding employing full-time human resources managers by obtaining advice and service through the Internet. A typical case is the API Group, a UK enterprise producing special packaging and security products used by consumer goods companies to help their products stand out on the shelf.

The success of the API Group is based on (a) the ability of the company to focus on what is important to the organization, which meant changes in infrastructure that involved moving the architecture of the information technology (IT) systems out to British Telecom (BT); (b) the company made Dell the standard supplier of computer hardware, making use of the existing support to do away with the need for support personnel on-site; (c) the implementation of Oracle on-Demand, which involves outsourcing hosting and management of hardware, software, and applications of IT systems.

The results of these changes have contributed to reducing the IT headcount, cut the cost of hiring and training IT support, and reduced IT operating costs by 40 percent. Perhaps the most important benefit for the API Group has been the fact that having the system supported by a remote hosting center has proved cheaper than having the company set up its own disaster recovery plan.

Other benefits of this approach are that users of the system enjoy higher service levels than before, making them more productive, and the updating of the system has contributed to an increase in effectiveness and efficiency as well as accommodating more users. This means that the changes allow the company to focus on its core manufacturing operations instead of worrying about IT.

Outsourcing as an innovation strategy helps the bottom line. However, one must realize that by outsourcing activities such as finance, human resources (HR), and IT, companies are able to gain access to higher levels of expertise than might be available in-house and also save costs. The classic example of these are call centers in India, the Philippines, and elsewhere in the world handling banking, accounting, engineering activities, and billing as well as customer service issues.

To be sure, anyone who deals with outsourcing must be aware of the risks. By far the most important concern should be whether or not the providers of all types of outsourcing are meeting the needs, costs, and controls of their customers. After all, if something goes wrong, it is the reputation of the company outsourcing rather than the service provider whose reputation is likely to be affected.

Outsourcing can cut costs, transfer responsibility, and allow a company to concentrate on its core business. But getting into bed with the wrong partner, especially internationally, carries risks. Often too many companies find themselves with mismatched expectations and deliverables that have fallen short.

To be successful, an organization must take a forward-thinking approach and question a supplier's service capabilities, check its references, and plan for contingencies. Here is a core list of what to watch out for:

1. Is the supplier an industry expert?
   a. Always check references.
   b. Find out whether the suppliers actually do the work or whether they simply farm it out to someone else while taking a premium on another service provider's work—you will be surprised to find out how often the latter happens.
2. Is the supplier legally able to provide the services?
   a. Make local inquiries to the extent possible.
   b. Obtain a valid confirmation from the supplier of their ability to provide the service. For example, in Sweden, if maintaining accounts is outsourced to a supplier outside Sweden, the tax office must provide prior consent. A supplier should have the expertise to inform you of this.
3. Have you identified all of your service needs to make sure that you are covered?
   a. Ask the supplier to help you to identify everything required in each country. Your supplier is meant to be the industry expert.
   b. Define services carefully; for example, employer payroll returns may seem an adequate definition of the service required, but it is much better to break it up as payroll withholding tax returns, fringe benefits tax returns, and stock option tax returns.
   c. Ensure flexibility in the terms of engagement to allow you to amend services as required.
4. What if the supplier fails to deliver to the required standards?
   a. Include a regular review agreement.

    b. Make sure that service levels, as defined, are measurable; for example, value-added tax (VAT) returns are ready one month before a deadline.

    c. Ensure that the supplier is liable to correct errors at its own cost; many do not!

    d. Include the right to terminate at your will with reasonable notice.

5. Does the supplier keep up to date with changes in regulations or technology?

    a. Check how the supplier keeps up with changes in technology or regulations. For example, if you were outsourcing your accounting and tax services in Australia, how would your supplier have made itself aware of the changes in taxation of stock options that occurred during 2008?

    b. Include terminology in the contract that commits the supplier to keeping up to date in their area of expertise.

6. What happens if your company undergoes a corporate reorganization or restructuring?

    a. Include clauses in the contract to deal with such changes, including the right to assign the contract to the new entity.

Finally, outsourcing is one more practice that organizations use to maximize their own business's strategic initiatives. The primary reasons for outsourcing are to save money and to free up time to spend on strategic business issues. Another reason why organizations (both manufacturing and non-manufacturing alike) consider outsourcing is to improve service quality. Key drivers for this improvement are access to outside expertise, the opportunity for cost savings, and a desire to focus resources on core business.

Although outsourcing is not new, what is new is the scale of outsourcing being done even in areas of nonmanufacturing. For example, it is not unusual to see organizations outsource engineering, janitorial, maintenance, and financial services. In some cases, even human resource departments are affected. Before outsourcing, the following issues have to be considered:

- Identify how outsourcing fits with strategic objectives.
- Agree on which core strategic competencies must be kept in-house and what can safely be outsourced.

- Identify a complete view of internal delivery costs, the main cost drivers, and the potential savings and investment.
- Develop a clear view of the capabilities and reputation of each of the main outsourcing providers.
- Standardize and simplify processes and procedures prior to considering outsourcing.
- Discuss in detail the concept of outsourcing with customers (employees and business managers) and other key stakeholders in the organization.
- Assess internal outsourcing and explore how shared services might deliver the same benefits but with greater retained control.
- Define the key success measures that will be used to judge the performance of the outsourced provider and the structure of the deal.
- Identify the technological challenges and solutions around outsourcing.
- Consider the history of the organization in terms of managing complex transition processes, including logistics, inventory, and scheduling.

SUPPLY CHAIN MANAGEMENT

We used to talk about vendors; then we moved to suppliers; and in the world of innovation, we finally realized that it is the supply chain that controls the overall quality of the organization. Operations and supply chain management are important in the widening of markets, more so under globalization. There is a need to serve the customer better at a lower cost. One needs to take care of primary, secondary, and reverse logistics in the execution of a business. The product, its quality, price, time, and place of delivery contribute to success in a competitive environment. Though production is limited by the capacity of an entrepreneur in relation to market demand, the right time and right place in meeting the customer in the distribution of the product have been the vital links in determining the measure of success.

One might be left with many choices in the execution of the task, but the ultimate success depends on the reduction of cost. This is where innovation and an understanding of supply chain management (SCM) come into play. SCM synchronizes production, scheduling, warehousing, transport, and other activities to create a holistic system for optimization of resources and profitability. The true innovation is in the recognition of SCM as a holistic system.

Effective supplier management is also likely to facilitate the building of strategic partnerships. Developing such partnerships with suppliers is vital to a successful supply chain. Whether called supplier relationship management, strategic partnerships, supply chain management, supplier rationalization, or strategic sourcing, activities preceding the signing of a contract that limit the number of suppliers an organization does business with are becoming a necessary way of doing business in today's competitive global marketplace.

The name may change, but establishing and maintaining a long-term relationship with suppliers will go a long way toward providing better communications and shared resources and rewards. Strategic supply management involves developing the strategies, approaches, and improvement from the procurement and sourcing process, particularly through direct involvement and interaction with suppliers (Monczka et al. 2009). The relationship is that of win–win for both organizations as they both reevaluate constantly both costs and improvement projects.

Companies have also started to limit the number of suppliers they do business with by implementing supplier review programs that identify suppliers with operational excellence. A close buyer–supplier relationship is important because suppliers in such a relationship are easier to work with and provide better service. An exemplary case is that of Ford Motor Company, which provides its supplier base with tremendous help for improvement through the supplier technical assistant engineer.

A close relationship obviously goes against the standard protocol of getting multiple quotes from vendors who may be cutting costs just to get a bid when business is slow only to surprise the buyer later with price increases or longer lead times.

COPING WITH THE COMPLEXITY OF SUPPLY CHAIN

There is a widely held belief among managers that recent years have seen a palpable increase in the risk of business disruption. This risk has many sources—not just the obvious and current threats from terrorism and geopolitical events but also the unexpected impact of particular business decisions. There is also strong evidence that the continued search for business efficiencies and the rise of what has been termed "the extended enterprise" have been the cause of a major change in the risk profile of many companies. This risk increasingly lies not within the organization itself but in the wider supply chain.

GROWING TURBULENCE AND UNCERTAINTY

In recent years, market turbulence has tended to increase for a number of reasons. Demand in almost every industrial sector seems to be more volatile than before. Product and technology life cycles have shortened, and competitive product introductions make life cycle demand difficult to predict. Considerable chaos exists in our supply chains through the effects of actions such as sales promotions and quarterly sales incentives. Sometimes this chaos can be exacerbated by the imposition of arbitrary rules, such as minimum order quantities or safety stock decisions.

At the same time, the vulnerability of supply chains to disturbance or disruption has increased. In 2003, the Gartner Group, a U.S.-based research company, predicted that one in five businesses would be impacted by some form of supply chain disruption and that, of those companies, 60 percent would go out of business (Christopher 2005). Furthermore, many companies have experienced a change in their supply chain risk profile as a result of changes in their business models. For example, the adoption of Six Sigma methodology, Lean practices, the move to outsourcing, and a general tendency to reduce the size of the supplier base can all potentially increase supply chain vulnerability.

The impact of unplanned and unforeseen events in supply chains can have severe financial effects across the network as a whole. Research in North America and reported in Christopher (2005) suggests that, when companies experience disruptions to their supply chains, the impact on their share price once the problem becomes public knowledge can be significant. The same research reported that companies experiencing these sorts of problems saw their average operating income drop by up to 107 percent, return on sales fall by 114 percent, and return on assets decrease by 93 percent year after year.

In 2002, Land Rover, part of the Ford Motor Company, announced that it might have to halt production of its Discovery four-wheel drive vehicle because its sole supplier of chassis—UPF-Thompson—had gone into liquidation. It was estimated that it could take up to 6 months for an alternative source of supply to be brought on stream. At significant cost, Land Rover had no alternative but to finance the supplier to enable the production of the chassis to continue.

Clearly, there are risks that are external to the supply chain and those that are internal. External risks may arise from natural disasters, wars, terrorism, epidemics, or government-imposed legal restrictions. *Internal risks* refers to the risks that arise as a result of how the supply chain is structured and managed. Though external risk cannot be influenced by managerial actions, internal risk can (Christopher 2005).

WHY ARE SUPPLY CHAINS MORE VULNERABLE?

A study conducted by Cranfield University for the UK Department for Transport and reported by Christopher (2005) identified a number of reasons why modern supply chains have become more vulnerable. These factors include the following.

A Focus on Efficiency Rather Than Effectiveness

The prevailing business model of the closing decades of the twentieth century was based upon the search for greater levels of efficiency in the supply chain. Experience highlighted that there was an opportunity in many sectors of industry to take out significant cost by focusing on inventory reduction. Just-in-time practices were widely adopted and organizations became increasingly dependent upon suppliers. This model, though undoubtedly of merit in stable market conditions, may become less viable as volatility of demand increases. The challenge in today's business environment is how best to combine Lean practices with an agile response.

The Globalization of Supply Chains

There has been a dramatic shift away from the predominantly "local for local" manufacturing and marketing strategy of the past. Now, through offshore sourcing, manufacturing, and assembly, supply chains extend from one side of the globe to the other. For example, components may be sourced in Taiwan, subassembled in Singapore, and assembled in the

United States for sale in world markets. The motivation for offshore sourcing and manufacturing is usually cost reduction. However, that definition of cost is typically limited to the cost of purchase or manufacture. Only rarely are total supply chain costs considered. The result of these cost-based decisions is often higher levels of risk as a result of extended lead times, greater buffer stocks, and potentially higher levels of obsolescence—particularly in short–life cycle markets. A further impetus to the globalization of supply chains has come from a consolidation of the supplier base as a result of the increase in cross-border mergers and acquisitions that we have witnessed over the past decade or so.

Focused Factories and Centralized Distribution

One of the impacts of the implementation and subsequent enlargement of the single market within the European Union and the consequent reduction in barriers to the flow of products across borders has been the centralization of production and distribution facilities. Significant scale economies can be achieved in manufacturing if greater volumes are produced on fewer sites. In some cases, companies have chosen to focus their factories—instead of producing the full range of products at each site, they produce fewer products exclusively at a single site. As a result, production costs may be lower, but the product has to travel greater distances, often across many borders. At the same time, flexibility may be lost because these focused factories tend to be designed to produce in very large batches to achieve maximum scale economies.

Along with the move to fewer production sites is the tendency to centralize distribution. Many fast-moving consumer goods manufacturers aim to serve the whole of the Western European market through a few distribution centers.

The Trend to Outsourcing

One widespread trend, observable over many years, has been the tendency to outsource activities that were previously conducted within the organization. No part of the value chain has been immune to this phenomenon; companies have outsourced a wide range of activities, including distribution, engineering, manufacturing, accounting, and information systems. In some cases, these companies might more accurately be described as virtual companies.

There is a strong logic behind this trend, based on the view that organizations are more likely to succeed if they focus on the activities in which they have a differential advantage over competitors. This is leading to the creation of network organizations, whereby confederations of companies are linked together—usually through shared information and aligned processes—to achieve greater overall competitiveness. However, outsourcing also brings with it a number of risks, not least the potential loss of control. Disruptions in supply can often be attributed to the failure of one of the links and nodes in the chain. By definition, the more complex the supply network, the more links there are; hence the greater the risk of disruption.

Reduction of the Supplier Base

A further prevailing trend over the past decade or so has been a dramatic reduction in the number of suppliers from which an organization typically will procure materials, components, and services. In some cases, this has been extended to single sourcing, whereby one supplier is responsible for the sole supply of an item. Several well-documented cases exist where major supply chain disruptions have been caused by a failure at a single source. Even though there are many benefits to supplier base reduction, this does involve increased risk.

Achieving Supply Chain Resilience

Because even the best-managed supply chains will hit unexpected turbulence or be affected by events that are impossible to forecast, it is critical that resilience be built into them. *Resilience* implies the ability of a system to return to its original or desired state after being disturbed. Resilient processes are flexible and agile and are able to change quickly. Supply chain resilience also requires slack at those critical points that can be adversely affected by changes in the rate of flow.

Supply chain resilience depends on rapid access to information about changed conditions. Through collaborative working with partners, this information can be converted into supply chain intelligence. Because networks have become more complex, they will rapidly descend into chaos unless they can be connected through shared information and knowledge. The aim is to create a supply chain community where there is a greater visibility of upstream and downstream risk and a shared commitment to mitigate and manage those risks.

Ultimately, it may be necessary for companies to reengineer their supply chains not, as in the past, with cost minimization in mind but to maximize their flexibility and agility. Today's changed conditions are forcing companies to question past decisions on sourcing, outsourcing, and the pursuit of Lean solutions. Responsiveness and resilience must be the twin goals of supply chain design and management.

SUMMARY

In this chapter we focused on innovation and elaborated on specific items of how the organization may benefit from them. Specifically, we discussed innovation in design, manufacturing, outsourcing, and supply chain. In the next chapter we will focus on people and quality.

SPECIAL NOTE

Some of the material in this chapter has been adapted from Martin Christopher's "Coping with Complexity and Chaos" and Mark Jones and Fran Samalionis' "From Small Ideas to Radical Service Innovation" with permission from the authors.

ENDNOTES

1. Zipcar, Inc., provides car sharing and car club services for individuals, businesses, and universities throughout the eastern United States, London, England, western Canada, and some western U.S. cities.
2. Google is the number one search engine.
3. A good example of innovative service is the case of the First Source Bank in South Bend, Indiana, which resulted in a brand new banking paradigm. It was the first bank to employ the side-by-side banking configuration. The change to the new paradigm resulted in new accounts, deposit growth, and loan sales, all more than 30 percent higher than for similar branches lacking the new service model and facility elements.

REFERENCES

Administration. (June 14, 2010). http://indiatelecomnews.com/?p=742. Downloaded August 29, 2011.

Bureau of Labor Statistics. (December 2005). http://www.bls.gov/ces/highlights122005.pdf.

Christopher, M. (2005). Coping with complexity and chaos. *Financial Times*. http://www.ft.com/intl/cms/s/1/28d3583c-25eb-11da-a4a7-00000e2511c8.html#axzz1YcLtO5yS

Global Employment Trends. (January 2007). a newsletter published by the International Labor Office. http://www.ilo.org/empelm/pubs/WCMS_114295/lang--en/index.htm

Gupta, P. (2007). *Business innovation for the 21st century*. New York: BookSurge. http://www.ft.com/intl/cms/s/1/28d3583c-25eb-11da-a4a7-00000e2511c8.html#axzz1S1auM4ri

Innovation. http://www.amanet.org/images/hri_innovation.pdf

Jones, M. and F. Samalionis. (2008). "Radical service innovation." *Design Management Review* October, 41–45.

Jones, M. and F. Samalionis. (winter 2008). "Radical Service Innovation." first published in Design Management Review Vol. 19 No. 1. And found in the web site of http://www.ideo.com/images/uploads/news/pdfs/08191JON20.pdf

Monczka, R., R. Handfield, L. Giunipero and J. Petterson. (2009). *Purchasing and Supply Chain Management*. 4th ed. Mason OH: South-Western Cengage Learning

Parker, P. (Ed.) (2006). *Webster's Online Dictionary*. Princeton, NJ: Princeton University.

SELECTED BIBLIOGRAPHY

Bartholomew, D. (2001). "Turbocharging the supply chain." *Industry Week* 250(12): 59.

Bowersox, D. J. and D. J. Closs. (1996). *Logistical management*. New York: McGraw-Hill Companies.

Christopher, M. (2005). "Coping with complexity and chaos." *Financial Times* 16 September.

Craumer, M. (2002). "How to think strategically about outsourcing." *Harvard Management Update,* May, 23–27.

Ford Motor Company. (2008). *Corporate quality development center (CQDC) training*. Dearborn, MI: Ford Motor Company.

Geiger, C., J. Honeyman, and F. Dooley. (March 1997). *Supply chain management: Implications for small and rural suppliers and manufacturers*. Fargo, ND: Upper Great Plains Transportation Institute.

Insinga, R. C. and M. J. Werle. (2000). "Linking outsourcing to business strategy." *Academy of Management Executive* November, 23–27.

Ketter, P. (2007). "HR outsourcing accelerates." *T&D* February, 12–13.

Kumar, S. and R. Bragg. (2003). "Managing supplier relationships." *Quality Progress* September, 24–30.

Pourier, C. C. and S. E. Reiter. (1996). *A new look at business partnering, supply chain optimization*. San Francisco: Berrett-Koehler Publishers.

Powell, A. S. (2002). "Transforming the supply chain." *The Conference Board,* March, 3–45.

Sawyer, C. A. (2007). "The innovation situation." *Automotive Design and Production* December, 54–55.

Sutton, R. (2001). *Weird ideas that work: 11 1/2 Practices for promoting, managing, and sustaining innovation*. New York: Free Press.

Williams, T. A. (2008). "Innovation from the ordinary." *Quality* November, 52.

5

People and Quality

In the last chapter we discussed some critical issues of innovation and how organizations may take advantage of them in terms of improvement. In this chapter we discuss the relationship between people and quality. Specifically, we address the issues of human understanding, empowerment, employee loyalty, multicultural management, ethics, and training.

The greatest asset of any organization is their people. As such, every organization should pay attention to who they hire and who they keep. This attention is a quality issue, and it must be directed from the top management ranks to the human resources (HR) department. It must be well-thought-out and executed with attention to detail for the overall benefit of the organization. This is where leadership comes into play. This means that leadership must recognize that the relationship between leadership, strategy, human behavior, decision making, and organizational systems are of paramount importance, especially when leaders deal with the human side of the organization. This is not an easy task for anyone. This requires the ability to lead and make sound business decisions that are very complex and require both knowledge and skills. Some fundamentals are discussed next.

UNDERSTAND THE HUMAN DIMENSION

Leaders must create and cultivate a long-term employee motivation for the improvement of the organization. That means that the appropriate and applicable rules must be in place for predicting human behavior in the organization, as well as developing a "defusing" dysfunctional behavior and managing it.

In today's world, management must realize that organizations have at least four generations working side by side as part of the workplace. As such, it is imperative that the effort is made to understand their needs and make them enthusiastic about the organization that they work for. Here are some generalizations that, though they may not accurately describe every baby boomer or Gen Xer in all organizations, they can be used as a rough generational guide. On the other hand, these generalizations may be used as an excuse to break the ice with a coworker who might share a fondness for Super Mario Brothers or attendance at Woodstock.

- *Traditionalists:* aka, the Silent Generation, veterans. Born: between about 1925 and 1946. Cultural influences: Great Depression, World War II, Korean War, postwar boom era, GI Bill. Workplace values: loyalty, recognition, hierarchy, resistance to change.
- *Baby Boomers:* aka, the Sandwich Generation (because many take care of both children and aging parents). Born: between about 1946 and 1964. Cultural influences: popularization of television, assassination of President John F. Kennedy, the Beatles, first moon walk, Vietnam War, antiwar protests, sexual revolution. Workplace values: dedication, face time, team spirit.
- *Generation X:* aka, the Slacker Generation, the Me Generation. Born: between about 1964 and 1982. Cultural influences: fall of the Soviet Union, women's liberation movement, MTV, grunge, rise of home video games and personal computers, birth of the Internet, dot-com boom and bust. Workplace values: work–life balance, autonomy, flexibility, informality.
- *Generation Y:* aka, Millennials. Born: between about 1982 and the late 1990s. Cultural influences: Internet era, September 11 terrorist attacks, cellphones, Columbine High School massacre, Facebook. Workplace values: feedback, recognition, fulfillment, advanced technology, fun.

ASSUMING THE HELM

Though leaders may not actually do the work, they must always show and demonstrate their commitment in managing the transition to new technology as well as human workgroups. This demonstration may take the form of developing and maintaining the support of that workgroup or provide the basis of a directive that is simple but effective to improve workgroup performance. Above all, they must have a strategy to accomplish the vision of the organization.

SHAPING CORPORATE CULTURE

It is imperative that the leaders of the organization understand the critical components of corporate culture and how the organizational norms form into productive workgroups. In addition to this understanding, they must provide a direction to defuse the dysfunctional culture (if it exists) and establish the rules and rubrics of a high-performance work environment.

CREATING ORGANIZATIONS THAT WORK

In case the organization is in deep trouble, the leader must begin with creating an organizational charter. This must be followed by selecting an effective management team that thinks systemically and understands the value and importance of managing image and expectations not only from the internal culture but the external culture (ultimate customers and financial institutions). With his team, the leader must develop and communicate meaningful performance indicators for the organization that everyone can understand and follow within reasonable deadlines. Furthermore, an excellent leader will provide an atmosphere in which turf battles and duplication are eliminated and therefore suboptimization will no longer exist.

FACILITATING STRATEGIC DECISIONS

Quality and leadership are very closely related. Whereas leadership has vision for the future, quality initiatives and strategies will help a leader with critical distinctions between problems, decisions, and polarities. In addition, principles of quality methodologies and tools will expedite the processes of: (a) how to properly frame a decision, (b) selecting the decision makers, (c) how to identify common errors in the decision-making process, (d) defining essential elements of the effective decision-making process, and (e) knowing when the decision has been made.

The leader's role in creating effective strategies is a core requirement of his or her existence as a leader. But what does it mean to create effective strategies? At the very minimum it means that leaders must begin contemplation of their vision with both internal and external considerations that affect the organization from growth, financial stability, and good community citizenship perspectives. This in turn means that they must anticipate the impact of adaptive responses in such a way that appropriate framing of the strategic initiative is considered, as well as possible barriers to the specific implementation strategy. So how can a leader lead the strategic change in an organization? The following steps are essential:

- Plan a change initiative. This includes gaining credibility in executive circles.
- Build networks and relationships within the organization. This will help in focusing on the change, as well as avoid derailing the objectives.
- Always keep in mind the political reality of the organization as opposed to being idealistic. This will help in being consistent and not dysfunctional in your objectives.
- Recognize that power can be gained as fast as it can be lost.
- Identify critical variables very early in the process of organizational change.
- Take an active role as the leader in fostering change.
- Plan and anticipate resistance and have contingency plans.
- Plan and anticipate for the culture to be a barrier to change and have appropriate contingency plans.

All organizations depend on people. Therefore, the first stage of engagement is recruitment followed by development and retention. The focus of recruitment is to hire individuals who have knowledge but also street smarts, commonsense, hustle, drive, and ambition. Development is the process of mentoring the employees to the next generation who will run the company. The focus of retention is to keep the recruits by taking care of them; they, in turn, take care of their customers, which in turn increases company revenues.

The essence of recruitment is to find, select, and offer employment to talented people. It is the talent and the person's knowledge that will bring future creativity in the organization. It is management's responsibility to unlock the power of creativity in all employees, because that creativity will turn ideas into actionable strategies for everyone.

Furthermore, allowing open participation will create loyalty to the organization, because everyone will sense some sort of ownership and personal satisfaction of having something to do with the decision-making process. Maximizing the creative power of the employees means success for undervalued, unique talents from the entire organization. Of course, the trick is to integrate these individual talents into a cohesive strategy that will bolster the company's values and goals.

The development stage is for making sure that employees are groomed for personal as well as professional improvement. Therefore, it is the responsibility of management to plan for individual success. That means that management must identify and have plans for the following:

- Steps to succession
- Creating a formal succession document
- Identifying high-potential employees
- Partnering candidates with mentors
- Broadening candidates' skill bases with rotations
- Conducting annual performance reviews
- Managing expectations
- Developing contingency plans
- Being prepared for defections
- Communicating often, precisely, and discreetly

The retention stage is for making sure that the best and brightest are kept within the organization. A recent study (Nancheria 2008) showed that 84 percent of employers with new or revised retention programs say that they have been successful in retaining high-potential employees due to the following:

- More careful selection, 63 percent
- Flexible work schedules, 42 percent
- Improved training, 61 percent
- Tuition reimbursement, 38 percent
- Coaching, 54 percent
- Exit interviews, 38 percent

- Better compensation and benefits, 52 percent
- Retention bonuses, 43 percent
- Better orientation, 51 percent
- Casual dress codes, 24 percent
- Mentoring, 43 percent
- Health insurance, 20 percent

EMPOWERING EMPLOYEES

Empowerment is a principle that is built on trust, respect, integrity, first-class service, innovation, responsibility, accountability, and the opportunity to have an input on idea generation as well as decision making and be part of a team environment (family culture). These characteristics/values work when they are transferred to one of the company's shops (lowest level operators).

Instead of giving managers rigid guidelines on how to run their business units, the empowering philosophy empowers them to embrace their entrepreneurial flair to personalize and give the proper authority and responsibility to the person who is responsible for a task.

Empowerment is challenging for everyone. It means overcoming the natural tendency of all management (in all organizations) to think that they know best. Otherwise, there is a temptation to talk about giving people power in theory while fearing to do so in each individual circumstance.

To avoid the path of least resistance, we need to fully adopt the subsidiary principle. That means that any organization that is attempting to practice empowerment must try to devolve power to the lowest possible level and then work upwards if that is not feasible. This would mean starting with the assumption that we can give power to the individual. Our default position should be to give them choice over what management provides for them. Typical examples may be in options for retirement plan, health insurance, problem-solving approaches to specific problems, and so on.

On the other hand, we should also recognize that individual choice does not always work. Where it does not we should escalate the power to the supervisor, then to the department head, plant manager, and so on. Executives and managers should do only that which only they can do. This means that executives will be focusing on long term strategies for the organization, rather than details of people's lives. This radical devolution is not just right in principle but also the most efficient way of creating a system that is adaptable enough to today's rapid change.

WORKER LOYALTY

For every organization, the skills of their employees have been and continue to be an important issue. This is particularly true in medical fields (especially nursing), construction, manufacturing, and service sectors of our economy where skilled openings are constantly advertised. The problem is exacerbated by poor retention policies. It has been estimated that on average one in five employees will quit their current employer for something better. When one realizes that the entire hiring process, on average, costs about 6 months' salary, it is clear that the direct cost to employers is considerable.

The true implications of these costs are far greater, with respect to productivity gaps between successful/unsuccessful organizations and in some cases between countries, such as France, Italy, Germany, and the UK. Yet, there is little evidence worldwide that businesses are doing enough to address staff retention, improve skill levels, and raise productivity.

The world has changed, and the expectations of employees are quite different from yesteryear. For example, gone are the days where employees worked for just one organization. Now, on average, each employee will work for four to five different employers in their careers. Even with this drastic social change, employers must realize and accept the fact that they must invest in people and ensure that human resource policies have a strategic role. (In the United States and in some EU countries the government is giving substantial funding to organizations that have ongoing, active training and skill improvement programs—some call this *corporate welfare*.)

If businesses are serious about productivity and long-term profitability, they need an effective people strategy that focuses on getting the right people to begin with. Retention of skilled staff can only be achieved if there are transparent HR strategies to ensure career progression, vocational training, and professional development opportunities meet the long-term needs of business and employee. In their exit interviews more often than not employees cite a lack of career progression second only to salary as their main reason for leaving.

A case in point is the notion of leaders and leadership. We know from many psychological studies that leaders are made, not born, suggesting that training and development plays a vital part in their shaping. There are, of course, exceptions to this. Key development courses could be communication, motivation skills, decision-making skills, evaluation of risks, and other specific topics (i.e., negotiation, selling, etc.) that would help a leader run his company effectively.

Businesses therefore need to review training opportunities for all employees, working hours, locations, childcare availability, flexible working hours, and other benefits and to be seen to be doing this through HR communication. The majority of organizations say—when asked—that they are focused on employee commitment, but only a few have actually direct staff communications, satisfaction surveys, and HR effectiveness benchmarking (especially for salary).

Of course, lack of review the training opportunities does not mean staff costs increase. On the other hand, when developing benefits or even buy-out packages, there are a host of potential options available, and companies can take advantage of all of them to create a positive environment for all concerned. It is indeed short-term thinking to think only how to manage payroll cost as part of the rationale in reviewing HR strategy.

MULTICULTURAL MANAGEMENT

With globalization and market liberalization, managers the world over are now having to contend with the fact that cultural competency is as vital as technical and management skills. Dealing with people from diverse cultures does not come naturally but requires thought, effort, and perseverance.

The borderless world we now live in has made it commonplace for managers to have to deal simultaneously with British and American business partners, Japanese suppliers, and Arab government officials, all involved in the same project. It is not unusual for something to be designed in the United States, manufactured in China, assembled in Brazil, and sold in Europe.

Defined as a set of norms that a group of people agree to follow in order to coexist, *culture* influences the ways in which we communicate and behave because the set of norms encompasses our thought patterns, beliefs, self-image, and emotional responses. When we talk about quality, differences in culture provide the opportunity to create synergy because diversity encourages creativity. Global success does not require one to start acting like an Arab, Chinese, Brazilian, European, Australian, or an

American; rather, success is achieved when one is able to, firstly, listen, watch, and feel accordingly when faced with different cultural norms and values. The next step is to allow one's self to react, participate, and grow. Only then will one be able to adapt, share, and experience the synergistic effects of multicultural dealings.

A successful global manager adapts to the values of other cultures while at the same time maintaining his or her own set of values. The successful manager in a global economy must master multicultural management and negotiation. That means that unless the manager has an open and receptive attitude when confronted with a new culture, he or she will not be successful. It is important to behave like a curious and inquisitive child, always asking "Why?" and "How?" so that you learn the reasons behind doing, or not doing, certain things, while at the same time avoiding making assumptions altogether.

It is imperative for any global manager to develop a habit of wanting to find out what went wrong—the cultural *faux pas* committed—should the deal fall through, because then there is less likelihood of the mistake being repeated in the future. For more effectiveness, managers should endeavor to recruit a mentor or a coach—one from the foreign culture—who will be able to assist and advice the manager.

Admittedly, even within the same culture, there are quite often different subcultures in which people value different norms and beliefs. However, it is still possible to conclude that, generally, individuals originating from the same culture tend to behave in a particular manner. Malaysians are a typical example of this. Generally, Malaysians and Singaporeans tend to value family, group harmony, cooperation, relationships, and spirituality, although there are multiracial groups in each country with differing cultural origins as well as faiths.

From the author's experience consulting, training, and teaching in over 58 countries on all continents, the following observations are given as a guide to global managers (this is not a complete list):

- *Americans:* They communicate directly and openly.
  - Be persistent but not overbearing.
  - Let them know how they stand to benefit from your products or services rather than attempt to awe them with the people you know.
  - Americans evaluate propositions from other companies on their technical merits and immediate benefits.

- While discussing business with Americans, allow participants to express their views openly.
- If an American company gives you an opportunity to do work for them, make sure that you do an exceptional job. American companies often evaluate your performance rigorously.
- *British:* Formalities and protocol are extremely important.
  - Meetings are established well in advance (so rule out any thoughts of making surprise visits) and have clearly defined purposes. Presentations should be detailed and easily understood.
  - The British tend to get down to business after a few moments of polite conversation.
  - Opinions are welcomed, but agreement may be slow.
  - Decision making can be adversarial.
  - Consensus is preferred to individual initiative, but groups are reluctant to take responsibility for error.
- *Germans:* Last names and appropriate titles are used until specifically invited to use first names. Academic and professional titles are also used frequently.
  - Germans value punctuality. Call or write with an explanation if you expect to be delayed.
  - Business meetings are formal and scheduled weeks in advance. The main purpose of any first meeting is to gain trust.
  - Germans come to meetings well prepared. Substantiate your presentations with facts, figures, and charts.
  - German corporate organizations are methodological and compartmentalized with procedures and routines done by the book.
  - Objective criticism is not easily given or accepted.
  - Compliments for a job well done are seldom given because Germans see it as your job to do a good job.
- *Japanese:* Establish a close relationship with a Japanese person who will serve as your sponsor. Make sure that your sponsor introduces you to a company contact who is mutually obliged to him.
  - They value the exchange of business cards upon first introductions, with priority given first to the company you are from, followed by your position and, finally, your name.
  - Japanese meet with foreigners to collect information, not to reach decisions.
  - Japanese prefer group consensus to individual decision making. Individuals do not appreciate being singled out to be

complimented on a job well done because group harmony is highly valued.

- Be aware of nonverbal cues given out by the Japanese because they give you an indication of the turn the meeting is taking. For example, a smile and nod of the head does not necessarily indicate agreement. Slouching or relaxing in the chair indicates boredom.
- *French:* Punctuality is expected. A quick and light handshake before and after meetings is the norm.
  - Use last names and titles unless invited to use first names.
  - Organizations are highly centralized and have powerful chief executives.
  - Work gets accomplished through a network of personal relationships and alliances.
  - Technical competency is admired as well as rivalry and competition.
  - The French get down to business quickly, but decisions are made only after deliberation.
  - Presentations should be made in a formal and professional style that appeals to the conservative business values of the French.
- *Chinese:* They will not discuss details immediately until they feel that you are committed long term.
  - Trust and mutual obligation are key factors in determining whether or not the Chinese will decide to do business with you.
  - If the policies of the Chinese government change significantly between signing and completion of the contract, they will ask to renegotiate.
  - Do not expect the Chinese to strictly adhere to the terms and conditions of a contract.
  - Decision making is based on consensus, involving people at several different levels of management.
- *Brazilians:* They will not discuss details right away. They like titles and prefer formal address while in a professional environment.
  - They are very team oriented.
  - They are very committed to quality.
  - They are eager to learn and apply new methodologies for improvement.
  - They are very detailed oriented.
  - They are more prone to formal presentations.
  - They are more likely to work in team projects.

- *Thais:* Very private. First meeting is to get to know each other.
  - Like teamwork.
  - They will agree with you, but that does not necessary mean agreement to the project. The final decision is made by the team and a supervisor after the meeting.
  - Work very hard in understanding the problem and offering a solution.
  - Like very active interaction as a team.
  - Very hard to decide on individual basis.
- *Taiwanese:* Establish a close relationship with a Taiwanese person who will serve as your sponsor.
  - They display friendliness early on.
  - They like very technical agreements.
  - They follow very strict instructions.
  - Decision making is based on consensus.
- *South Koreans:* Hard to read their posture and expectations.
  - Very quality oriented and willing to learn new quality methodology and/or specific tool for improvement.
  - Like to work in teams.
  - Decision making is based on consensus.
- *Swedish:* Very serious and objective oriented.
  - They like procedural approaches.
  - Use a lot of intuitive approaches.
  - Like to challenge the discussion points.
  - Do not like criticism.
  - They like praising.
- *Egyptians:* Very formal in their approach. They like titles, especially academic credentials.
  - Very cooperative.
  - Easy to communicate agreements and disagreements.
  - Eager to learn new ways of doing business.
  - Like formal agreements but will work without them if necessary.

BUSINESS ETHICS

Behaving unethically is bad for business. Worcster has pointed out (2007) that we do not want to go back to the days of the robber barons, when the only thing that mattered was profits, when buildings collapsed due to insufficient concrete; children were exploited as cheap labor; and sweatshops made hats from the feathers of endangered birds. The clichés that the only business of business is business, and what's good for General Motors is good for the country, have worn very thin indeed.

The corporate scandals that have hit the headlines recently were all due to unethical business behavior. These scandals are bad news for staff and shareholders and also for the businesses themselves. Enron paid the ultimate price of corporate collapse when its executives were jailed. In the process, it brought down not only its American auditors Arthur Andersen but also its associated British partners, innocent bystanders to their former partner's breach not only of their professional responsibilities but of company law.

The Institute of Business Ethics (IBE; 2003) was set up just over 25 years ago, in 1986: the year of the Guinness scandal in the UK and Ivan Boesky in the United States. You might think that not much has changed in those 20 years. But you would be wrong. Thanks to the IBE and many in business, the ethical behavior of business has shot up the boardroom agenda. Management and shareholders alike know that a business that is not run ethically is likely to be a bad business proposition. Indeed, in 2003 the IBE published a report showing a clear link between business ethics and financial performance.

What have businesses been doing over the years to improve their ethical behavior? First, they have been putting in place codes of ethics. In 1986, an IBE survey showed that only 18 percent of larger companies had codes of ethics; by 2006, more than 90 percent of Financial Times and Stock Exchange (FTSE) 100 companies had one. But a code of ethics, worthy as it may sound, is only as good if everyone adheres to it. After all, Enron had a very good code of ethics, but it was ignored by executives who exploited their positions. The same thing happened in the fall of 2008 when the financial institutions started to collapse. They were checked neither by their own code of ethics nor the law of the land, and they escaped scrutiny from their auditors, their boards, and their nonexecutive directors (Worcester 2007).

Much thinking in recent years is about how to embed a code of ethics within an organization to ensure that it becomes part of the corporate DNA, rather than a tedious document to be left in a drawer collecting dust. A great many more companies and their advisers are working on better ways of embedding ethical behavior within their organizations.

And, of course, part of that is not just about embedding it but also about monitoring and policing the codes of ethics. This involves every single person in a business; every employee must feel enabled to speak up about any instance of unethical behavior they may witness, and they must know that something is going to be done about it.

Improving ethical behavior is not just good for particular businesses but for the whole business community in general (Worcester 2007).

In 2007, an Ipsos Mori poll showed that over the past 3 years there was a significant rise in the number of people who believe that British businesses behave ethically.

Worcester also reports that in 2003, fewer than half of adults (48 percent) believed that British business behaved very or fairly ethically (2007). In 2006, that figure had risen to 58 percent. This has to be good for every stakeholder, employee, customer, and supplier. But as the all-too-frequent corporate scandals show, we still have some way to go to ensure that business ethics are an integral part of the life of every business.

A key issue for the next generation of business leaders will be to improve the ways in which ethics are managed within their business. It is to them that we must look for fresh thinking and for answers on how existing ideas might be applied as new issues arise. For example, will the rise in private equity ownership of some of our larger companies result in less transparency and, therefore, less concern about ethical behavior? Business ethics is now more important than ever. But it is also clear that this area will become ever more important in the future.

TRAINING

Workplace learning and performance professionals understand that culture can easily limit much of what we need to do. But because culture is hard to pin down in practical terms, let alone to effectively change for the better, it remains a baffling issue. However, organizational culture is simpler than our personal cultures, and it is much easier to change than we imagine.

Organizational cultures lack the broad links that help define how we understand ourselves, among others. This weakness also implies that organizational cultures are dynamic. The good news is that organizational cultures can adapt and change to new influences quickly.

Stories help define an organization's culture. It is easy to use stories to change that culture—simply get people to tell stories that amplify the best aspects of the organization. More important, tell positive stories— successful stories, things that have worked very well—to drown out the sound of competing stories.

The best way to get people to share good stories is through a cycle of inquiry, engagement, and review, in the context of training or rotation. If your organization follows this cycle, it will undoubtedly create culture change. More important, management will definitely be surprised at how effective, productive, content, and committed employees become and how much better it is to work at your organization.

In a knowledge economy, an organization's ability to quickly adapt to changing realities is critical to its success. To facilitate the upkeep of knowledge and skills, workplace learning professionals seek innovative training design models and delivery methods so that they can provide the right information at the right time to the right people. Of course, every learning solution is initiated for a specific reason. However, even with a specific reason, problems may arise because learning professionals typi- cally rely on qualitative evidence to build the business case for training.

When considering a significant investment or deciding among multiple requests, executives need to be able to review a quantitative measurement that addresses how training will help the unit and organization attain its goals; whether training is worthwhile; and how training compares to other organizational initiatives. For example, if the executive office is con- sidering 20 programs but can only fund 10, which ones should they select and why?

Learning professionals need to recognize that an assessment must occur during the planning stages when budgets and resources are allo- cated. In other words, training cannot rely solely on current evaluation models, such as Kirkpatrick's four levels, which assess training's impact after it has been delivered (2006). By then, it may be too late to calculate valuable results.

The need for rapid talent development has emerged as one of the major issues facing organizations in virtually all sectors of the economy. Traditional training is one way but certainly not the only one. Rotation is a second way of training. Rotation is a process by which high perform- ers rotate through different jobs inside and outside of different depart- ments and/or roles to broaden their perspective and expand their skill sets. Rotations also help executives assess whether their instincts about a particular employee are right as well as identify shortcomings and improvements for the candidate. Yet a third approach is e-learning. E-learning is a newcomer in the process and offers interesting opportunities for meeting this challenge. But before workplace learning and performance professionals adopt this widely used technology, there are some fundamental principles that need to be understood.

There is a simple framework for strategically thinking about new ways to support your talent development initiatives, and this framework is positioned around the acronym AIM (alignment, interactivity, and motivation), which directly targets blended e-learning programs to help you deliver the results you want at the speed you want (Adams 2008).

New win–win learning solutions that meet organizational and employee needs are essential. If there has ever been a time when technology is needed, this is it. However, this technology has not—as yet—been materialized to its full potential.

Adams (2008) claimed that we have the technologies. We just need to employ them using the AIM model to create a solid platform upon which to advance learning goals and meet the demands for rapid talent development.

What makes an employee want to stay at his or her job, especially a customer service employee, a star performer, or both? That question is one that many employers are being forced to examine. A recent study on popular retention methods from ClearRock, an outplacement and executive coaching firm in Boston, revealed that employers are having a tough time keeping both their frontline and high-potential employees—employees who are crucial to a company's success (Nancherla 2008).

In a ClearRock survey of 94 organizations nationwide, 37 percent of employers reported an increase in the turnover of frontline employees, and in the same time period, 31 percent of employers reported an increase in the turnover of high-potential employees (Nancherla 2008).

According to the survey, operations and production workers are the most difficult frontline employees to retain, followed by information services and computer-related workers, sales and marketing employees, and customer service employees.

High-potential operations and production workers are also the most difficult to retain, followed by workers in sales and marketing, information services, and accounting and finance. The results showed that 51 percent of employers instituted revised or new retention programs for their frontline employees; 56 percent of employers did the same thing for high-potential employees. However, employers with new or revised retention programs generally had more success retaining high-potential employees than front-line employees, with an 84 percent success rate for high-potential employees compared to 81 percent for frontline employees.

With the cost of replacing workers who leave or do not work out rising to two or three times their compensation, companies are revising their retention programs to make workers' tenures longer. Some of the top methods that employers are using in their retention programs include more careful selection, better compensation and benefits, better orientation and assimilation programs, coaching, the use of exit interviews, and improved training.

We must underscore the importance of not only the high-potential future leaders but also entry-level and customer-contact workers due to the effect of turnover on quality and customer service.

Training is typically viewed as something done apart from a job setting, perhaps in a classroom or lab area. You learn in training settings and you work in job settings. However, there is a fine distinction between on-the-job and training. Clearly, learning may be supported both on the job as well as in formal training settings. Conversely, aspects of the job environment can be successfully simulated in learning settings. A comprehensive treatment of training design would necessarily include on-the-job performance support in detail. For example, the traditional view of work settings tells us that work is done; people apply their knowledge by performing tasks and solving problems. On the other hand, a more inclusive interpretation may be that work is done with appropriate tools, info sources, job performance aids, help and advisement systems; workers continue to learn on the job through mentoring, apprenticeships, internships, and so on.

We also have observed that in the traditional learning setting, people learn in classrooms; there, they acquire the knowledge and skills needed to perform successfully on the job. Conversely, a more inclusive interpretation may be that people learn by introducing elements of the work setting—tools, aids, and help systems—into manageable and low-risk training environments. Job demands are simulated in controlled training settings.

In fact, this is the reason why most organizations believe in the rotation method—especially for their high-potential employees. But rotation alone is not enough; therefore, many companies are hurting and will continue to hurt unless something changes in their training schemes

Robert Half Management Resources has reported in one of their studies that some 83 percent of CFOs say that they have not identified a successor (quoted in O'Sullivan 2007). The primary reason? Most believe that they will hold the position for some time. With CFO tenure hovering at 3 to 5 years on average, however, that may not be realistic. It is also shortsighted, not to mention that it ignores the basic concept and/or rule of management, which is that the function of management is to do the job through others and to train your successor.

Succession planning takes time and training, which both are in short supply in most modern organizations. Singling out individuals to groom for the top job can create resentment among those not chosen, increasing the risk that they will walk out the door. Moreover, shifting star players from one position to another for training can disrupt the flow of work in both the department the star left and the one that he entered. A classic example of resentment is the situation at GE when Jack Welch retired. Three of his top reports were widely known to be in the running for his job. Jeffrey Immelt got the nod; within days, the other two candidates quit.

The person in the job is often not interested in talking about succession or training, because he or she wants to stay there for a long time. After all, succession planning does not affect the current quarter; therefore, it easily slips off any leader's immediate to-do list.

For a leader in any organization to see succession and training as a viable option within their objectives, both must be formalized. The first step involves encouraging executives to think about what would happen to their companies if they could not do their jobs for some reason. People then start thinking about the good of the organization as a whole rather than the protection of their own position. (We forget that no one is perfect and no one is irreplaceable.) The second step is for the executives to meet with senior teams several times a year to discuss the highest potential candidates in every function of the organization. Then these individuals are paired with mentors and considered for different career development possibilities, including assignments to new roles.

Commitment to communication helps retain talent and defuse the tension that often accompanies succession planning. But such communication has to be carefully managed. You need to communicate not only with the people who have been identified for rotations but also with those who have not been included. This communication preferably has to be conducted in one-on-one discussions between managers and their direct reports to set expectations, identify development needs, and make sure that there are no mismatches between expectations and reality. If mismatches are discovered, it is probably best if both sides admit that and move on.

Face-to-face meetings also help separate the more ambitious employees from the staffers who are content where they are. But many employees know which of their coworkers are stars and consciously or unconsciously defer to them already.

Conventional wisdom says that training is more context and job specific than education; training is skills based, whereas education is knowledge based. Another way of saying this is that training focuses on how, whereas education focuses on why. Unfortunately, this distinction has been used as an excuse for rote procedural training for many years. We believe that the distinction between training and education is not as clear-cut as many believe; when learning takes place, both knowledge and skills are acquired. Indeed, most training situations call for some degree of meaningful understanding and problem-solving capability.

Fortunately, the discipline of instructional design (ID; the science of developing and presenting material to the target audience) comes to the rescue. Its foundation rests on the following:

- A systems design model for managing the instructional development process.
- Theories that specify what high-quality instruction should look like (Reigeluth 1983, 1987; Stamatis 1986, 2003; and many others).

These foundations have served designers over the last 25 years, in public schools, higher institutions of learning, military, and industry. In fact, if one looks at the published literature, one will find that the models of instructional design deliver on their promise of effective instruction.

Of course, just like every other discipline, ID is evolving, with new models and methodological advances, such as systems thinking and project management (Tripp and Bichelmeyer 1990; for a full application of ID processes in a Six Sigma methodology, see Stamatis 2003). In addition, the discipline at large is using sophisticated computer-based tools to automate the ID process (Wilson and Jonassen 1990–1991).

ID is a dynamic discipline, and it continuously develops/modifies new/existing models for developing the best possible material/media for understanding. Typical methods/technologies are discussed next.

RAPID PROTOTYPING

The terminology and form are borrowed from the computer science system design principles. It is applied to ID to allow greater flexibility in defining the goals and form of instruction at early stages (Tripp and Bichelmeyer 1990; Wilson et al. 1993). Prototypes may be shallow or narrow: shallow in the sense that the entire look of a product is replicated minus some functionality or narrow in the sense that a small segment is completed with all functionality, leaving other whole portions of the final product undeveloped.

Prototyping can be relevant to all kinds of training development projects, but its value is most apparent in the design of computer-based systems (Wilson et al. 1993a). Rapid prototyping may be done for a variety of reasons, including to

1. Test out a user interface.
2. Test the database structure and flow of information in a training system.
3. Test the effectiveness and appeal of a particular instructional strategy.
4. Develop a model case or practice exercise that can serve as a template for others.
5. Give clients and sponsors a more concrete model of the intended instructional product.
6. Get user feedback and reactions to two competing approaches.

Rapid prototyping can help designers break out of the linear approach to design. Tripp and Bichelmeyer (1990) also argued that rapid prototyping is more in line with how people actually solve problems in other domains, which is far from a linear process.

In the area of automated ID systems we find that computers do indeed make the process more efficient and flexible in the following ways:

1. *Data management.* Bunderson and colleagues (1981) described ID as a loop that begins with analysis of expert performance and ends with learners demonstrating that same expertise.
2. *Task support.* A wide variety of production tasks can be supported by computers, ranging from graphics production to word processing to communication among team members.

3. *Decision support.* Computers can assist in many design decisions by providing aids such as (a) ready access to information, (b) checklists, (c) templates, (d) expert system advisors, and others.

Finally, developing a robust training program is a delicate and time-consuming process. But companies with the resources and the will to implement a plan see a real payoff, not only in the orderly transition and avoidance of outside search costs but also in the retention of their top talent.

FORMATIVE EVALUATION

Recall that systems theory requires constant self-monitoring and adjustment of the system. This is similar to the scientific method, in that we formulate hypotheses (designs) and test them, thereby supporting or altering our expectations. Formative evaluation is the primary means of doing this self-testing; at various stages, designers may try out instructional materials to improve their effectiveness. Formative evaluation may be performed in phases, beginning with expert review, one-on-one or small-group trials, and tryouts with the target audience under the conditions that the materials were designed to function in. Significant changes may occur as a result of these reviews from both designers and learners; however, whether changes occur or not, the reviews will produce a much better product than if the reviews did not take place (Wilson et al. 1993).

New methods and approaches to formative evaluation are based on cognitive assumptions of performance. Whereas a system's evaluation in the past tended to focus on learners' success in performing the criterion task, cognitive techniques seek to uncover thinking processes as they interact with the material. For instance, think-aloud protocols (Smith and Wedman 1988) require learners to think out loud as they work through instruction. Their verbal reports become a source of data for making inferences about their actual thinking processes, which in turn provide evidence concerning the effectiveness of instruction. Learners might also be asked to elaborate on their verbal reports, particularly the reasons for decisions that result in errors.

Computer-based instruction allows designers to collect records of learners' every action. This is particularly useful for systems with extensive learner control; for example, an audit trail of a learner's path through a hypertext lesson can suggest to designers ways to revise the lesson to make desired options more attractive. Audit trails consist of a report of all of the learners' responses while navigating through an information system, including the screens visited, the length of time spent interacting with parts of the program, choices made, information input into the system, and so on. Their primary purpose or function has been to provide evidence for formative evaluation of materials (Misanchuk and Schwier 1991), such as the paths selected by learners and where and why errors were committed. Designers can evaluate the frequencies and proportions of nodes visited, the proportion of learners electing options, and so on. This information can provide valuable insights into the patterns of logic or strategies that learners use while working through instructional materials. A primary advantage of such data collection, when compared with other formative evaluation techniques, is that it is unobtrusive, accurate, and easy to obtain, because it is automated (Rice and Borgman 1983).

Another method, constructive interactions, observes how pairs of learners use instructional materials together, making collaborative decisions about how to proceed through materials. The transcribed dialog provides evidence about learners' hypotheses and reasoning while interacting with instructional materials (O'Malley, Draper, and Riley 1985; Miyake 1986). These techniques, like the think-aloud protocols, have provided designers with a richer understanding of how learners are going about learning—not just whether they fail or succeed.

TRAINING AS A SUBSYSTEM

Systems theory views instruction as a system with many parts but also as a subsystem within layers of larger systems. To be successful, then, designers must carefully consider the context or situation in which instruction is being developed. Two methods intended to ensure a proper fit between instruction and its larger context are performance analysis and environment analysis.

Performance analysis is a specific, performance-based needs assessment technique that precedes any design or development activities by analyzing the performance problems of a work organization (Gilbert 1978; Harless 1979; Jonassen 1989; Wilson et al. 1993). It begins with the assumption that "if it ain't broke, don't fix it"; that is, if no problems have occurred, no solutions are needed. Assuming that a problem in the way employees are performing has been identified, performance analysis begins by investigating the cause of the performance problem. Problems can result from a lack of skills or knowledge, lack of motivation, or an inadequate environment (physical, administrative, or logistical) in which to perform. Solutions to each of these types of problems differ. The premise of performance analysis is simple: The solution needs to fit the problem. Motivational problems may be solved by changing the work incentives or by methods aimed at increasing workers' cooperation, feelings of ownership, confidence in their ability to perform, or valuation of the tasks. Environmental problems may be solved by reengineering the environment. Formal training is recommended only if the problem results from a lack of knowledge or skills. Training is seen as a last-resort remedy because of its relatively high cost.

Assuming that there is a verified learning need, can the instruction be successfully implemented in the context for which it is intended? Environment analysis investigates the context of any instructional system, both where the instruction will occur and how the instructional materials will be used (Tessmer 1990). Physical factors include the facilities and equipment available, the life span of the instruction, distribution of the sites, the management and coordination available, and even the physical climate. Use factors include descriptions of how and why materials will be used, characteristics of the learners who will use them, characteristics of the administrators of the materials, and important logistical factors such as production, storage and delivery services, and the resources available to support their use. Careful attention to these factors may seem tedious, yet any one of these factors can mitigate the effectiveness of the best designed instructional programs.

Today, training must be viewed from the perspective of *effectiveness*—solving the customer's problem, also known as the *power* of the help. Did the training work? How did the training affect the bottom line? Is there a substantial and documented improvement because of the training? Relating to the effectiveness are also issues of (a) availability of help (is there someone around who can answer questions regarding the problem that is being solved?), (b) relevance to the problem at hand (is the training trying to solve an immediate problem?), (c) understandability (is the help clear to the user?), and (d) efficiency (how is the timeliness and affordability regarding the help that the training provides?) (Ankeny, 2008).

Finally, each time training is offered, there is by definition some problem to be solved, some goal to be reached. Even though not all training problems are as neatly quantifiable (for example, we want to institute a new training program to reduce safety accidents), there are clear implications for training design that reinforce the lessons learned. Training is most effective when it

- Responds to an immediate performance need.
- Seeks to meet specific teaching moments with relevant, clear instructional messages and practice opportunities.
- Does not give too much or too little help.
- Does not get in the learners' way; rather, it:
  - fosters a learning culture.
  - motivates learners.
  - makes training problem centered.
  - helps learners assume control of their learning.
  - provides meaningful practice.

SUMMARY

In this chapter we discussed some of the key issues one may find in any organization. The focus was on people and quality and specifically the concerns that institutions and organizations have given the different human dimensions, empowerment, loyalty, multicultural management, ethics, and training. In the next chapter we will discuss time management and how we should all try to optimize it for better results.

SPECIAL NOTE

In this chapter we have adapted material from "Cognitive Approaches to Instructional Design" by Brent Wilson, David Jonassen, and Peggy Cole, with the permission of the authors, as well as from Charles M. Reigeluth's *Instructional-Design Theories an Models: An Overview of Their Current Status*, and the Dean H. Stamatis's "The Effects of Hierarchical and Elaboration Sequencing on Achievement in an Adult Technical Training Program."

REFERENCES

Adams, J. (2008). "Rapid talent development." *T&D* March, 68–72.

Ankeny, R. (2008). "Defensive training." *Crain's Detroit Business,* 2 September (p. 4).

Bunderson, C. V., A. S. Gibbons, J. B. Olsen, and G. P. Kearsley. (1981). "Work models: Beyond instructional objectives." *Instructional Science* 10: 205–215.

Gilbert, T. (1978). *Human competence: Engineering worthy performance.* New York: McGraw-Hill.

Harless, J. (1979). *Guide to front end analysis.* Newnan, GA: Harless Performance Guild.

Institute of Business Ethics (2003). http://www.ibe.org.uk/index.asp?upid=71&msid=12

Ipso survey 2007. http://www.ibe.org.uk/userfiles/briefing_4.pdf

Jonassen, D. H. (1989). "Performance analysis." *Performance and Instruction* 4: 15–23.

Kirkpatrick, D. (2006). *Evaluating Training Programs.* 3rd ed. San Francisco: Berrett-Koehler Publishers.

Misanchuk, E. R. and R. Schwier. (1991). Interactive media audit trails: Approaches and issues. Paper presented at the annual meeting of the Association for Educational Communications and Technology, 13–17 February, Orlando, Florida.

Miyake, N. (1986). "Constructive interaction and the iterative process of understanding." *Cognitive Science* 10: 151–177.

Nancheria, A. (2008). "Retention tension: Keeping high-potential employees." *T&D* March, 18.

Nancherla, A. (March 2008). "Retention Tension: Keeping High-Potential Employees." http://findarticles.com/p/articles/mi_m4467/is_200803/ai_n25420172/

O'Malley, C. E., S. W. Draper, and M. S. Riley. (1985). "Constructive interaction: A method for studying human–computer interaction." In *Human–computer interaction,* edited by B. Shakel. Amsterdam: North-Holland.

O'Sullivan, K. (November 1, 2007). Who's Next? Succession planning should be a critical exercise in finance. Too bad so many companies avoid it. http://www.cfonet.com/article.cfm/10023288/c_10085936?f=insidecfoeurope

Reigeluth, C. M. (Ed.). (1983). *Instructional-design theories and models: An overview of their current status.* Hillsdale, NJ: Lawrence Erlbaum.

Reigeluth, C. M. (Ed.). (1987). *Instructional-design theories in action: Lessons illustrating selected theories and models.* Hillsdale, NJ: Lawrence Erlbaum.

Rice, R. and C. L. Borgman. (1983). "The use of computer-monitored data in information science and communication research." *Journal of the American Society for Information Science* 34(4): 247–256.

Smith, P. L. and J. F. Wedman. (1988). "Read-think-aloud protocols: A new data-source for formative evaluation." *Performance Improvement Quarterly* 1(2): 13–22.

Stamatis, D. H. (1986). The effects of hierarchical and elaboration sequencing on achievement in an adult technical training program. Ph.D. Dissertation, Wayne State University.

Stamatis, D. H. (2003). *Six Sigma: Implementation.* Boca Raton, FL: St. Lucie Press.

Tessmer, M. (1990). "Environment analysis: A neglected stage of instructional design." *Educational Technology: Research and Development* 38(1): 55–64.

Tripp, S. D. and B. Bichelmeyer. (1990). "Rapid prototyping: An alternative Öðretim Tasarýmý strategy." *Educational Technology Research and Development* 38(1): 31–44.

Wilson, B. G. and D. H. Jonassen. (1990–1991). "Automated instructional systems design: A review of prototype systems." *Journal of Artificial Intelligence in Education* 2(2): 17–30.

Wilson, B. G., D. H. Jonassen, and P. Cole. (1993). "Cognitive approaches to instructional design." In *The ASTD handbook of instructional technology*, edited by G. M. Piskurich. New York: McGraw-Hill.

Wilson, B. G., Jonassen, D. H., & Cole, P. (1993a). Cognitive approaches to instructional design. In G. M. Piskurich (Ed.), *The ASTD Handbook of Instructional Technology* (pp. 21.1–21.22). New York: McGraw-Hill.

Wilson, B. G., Jonassen, D. H., & Cole, P. (1993). ttp://carbon.ucdenver.edu/~bwilson/training.html

Worcester, B. (May 17, 2007). Why an ethical business is more likely to produce a success story. http://www.telegraph.co.uk/finance/2809088/Why-an-ethical-business-is-more-likely-to-produce-a-success-story.html

SELECTED BIBLIOGRAPHY

Baber, A. and L. Waymon. (2008). "Uncovering the unconnected employee." *T&D* May, 60–69.

Bednar, A. K., D. Cunningham, T. M. Duffy, and J. D. Perry. (1991). "Theory into practice: How do we link?" In *Instructional technology: Past, present, and future*, edited by G. Anglin. Englewood, CO: Libraries Unlimited.

Bransford, J. D. and N. J. Vye. (1989). "A perspective on cognitive research and its implications for instruction." In *Toward the thinking curriculum: Current cognitive research. 1989 ASCD Yearbook*, edited by L. B. Resnick and L. E. Klopfer. Alexandria, VA: Association for Supervision and Curriculum Development.

Brown, J. S., A. Collins, and P. Duguid. (1989). "Situated cognition and the culture of learning." *Educational Researcher* January–February, 32–42.

Brown, J. S. and K. VanLehn. (1980). "Repair theory: A generative theory of bugs in procedural skills." *Cognitive Science* 4: 379–426.

Gagne, R. M. and M. D. Merrill. (1990). "Integrative goals for instructional design." *Educational Technology Research and Development* 38(1): 22–30.

Galagan, P. (2008). "Talent management: What is it, who owns it, and why should you care?" *T&D* May, 40–51.

Gery, G. J. (1991). *Electronic performance support systems*. Boston: Weingarten Publications.

Graf, D. (1991). "A model for instructional design case materials." *Educational Technology Research and Development* 39(2): 81–88.

Keter, P. (2008). "What can training do for Brown?" *T&D* May, 30–39.

Maister, D. H. (2008). "Why (most) training is useless." *T&D* May, 52–59.

Mayer, R. E. (1990). *The promise of cognitive psychology*. Lanham, MD: University Press of America.

Nickerson, R. S. (1991). "A minimalist approach to the 'paradox of sense making.' A review of J. M. Carroll, The Nurnberg funnel: Designing minimalist instruction for practical computer skills." *Educational Researcher* December, 24–26.

Pace, A. (2008). "Talent management 2020." *T&D* December, 38–39.

Resnick, L. B. (1987). "Learning in school and out." *Educational Researcher* December, 13–20.

Rossett, A. (1987). *Training needs assessment*. Englewood Cliffs, NJ: Educational Technology Publications.

Schon, D. (1987). *Educating the reflective practitioner: Toward a new design for teaching and learning in the professions*. San Francisco: Jossey-Bass.

Taylor, R. (1991). "NSPI president challenges instructional design profs." *ITED Newsletter* 1: 4–5.

Tessmer, M., D. Jonassen, and D. C. Caverly. (1989). *A nonprogrammer's guide to designing instruction for microcomputers*. Englewood, CO: Libraries Unlimited.

Tessmer, M. and J. F. Wedman. (1990). "A layers-of-necessity instructional development model." *Educational Technology Research and Development* 38(2): 77–86.

Whitehead, A. N. (1929). *The aims of education*. New York: MacMillan.

Whitten, J. L., L. D. Bentley, and V. M. Barlow. (1989). *Systems analysis and design models*, 2nd ed. Homewood, IL: Irwin.

Winn, W. (1990). "Some implications of cognitive theory for instructional design." *Instructional Science* 19(1): 53–69.

6

Time Management

In the last chapter we discussed some issues with people and quality and how they may play a role in the improvement process. In this chapter we address the issue of time management. Specifically, we will address how time management skills affect people in their accomplishment of a task.

Perhaps one of the most underutilized tools in modern quality is the tool/activity/issue of time management. Time management for the individual is what project management is to big projects. Both are important for efficiency in accomplishing a task. Personal time management skills are essential skills for effective people. People who use these techniques routinely are the highest achievers in all walks of life, from business to sports to public service. In the twenty-first century, with exponential demands on our time, we all must be competent of these skills well, to be able to function exceptionally well, even under intense pressure. In addition, as we master time management, we will find that we take control of our workload and say goodbye to the often intense stress of work overload.

At the heart of time management is an important shift in focus: Concentrate on results, not on being busy. That means: learn to prioritize.

Many people spend their days in a frenzy of activity but achieve very little, because they are not concentrating their effort on the things that matter most. One way to learn how to shift priorities is to utilize the Pareto principle or, as it is commonly known, the 80:20 rule. The rule says that typically 80 percent of unfocused effort generates only 20 percent of results. This means that the remaining 80 percent of results are achieved with only 20 percent of the effort. Though the ratio is not always 80:20, this broad pattern of a small proportion of activity generating nonscalar returns recurs so frequently that it is the norm in many situations.

By applying time management tips and skills (for selected few see Appendix F on the companion CD-ROM), you can optimize your effort to ensure that you concentrate as much of your time and energy as possible on the high-payoff tasks. This ensures that you achieve the greatest benefit possible with the limited amount of time available to you. However, overall the advantages of time management may be summarized as:

- Gains time
- Motivates and initiates
- Reduces avoidance
- Promotes review
- Eliminates cramming
- Reduces anxiety

To reap these benefits, there are two fundamental keys that need to be understood:

1. *Self-knowledge and goals:* In order to manage your time successfully, having an awareness of what your goals are will assist you in prioritizing your activities.
2. *Developing and maintaining a personal, flexible schedule:* Time management provides you with the opportunity to create a schedule that works for you, not for others. This personal attention gives you the flexibility to include the things that are most important to you.

What does it take to have a quality time management attitude and plan? By far the most important is a personal schedule that works for you. It should include the following items:

A master schedule (for long-term planning)
A personal schedule (for short-term planning)

In both cases, the focus should be on:

- *Goal Setting:* To start managing time effectively, you need to set goals. When you know where you are going, you can then figure out what exactly needs to be done, in what order. Without proper goal setting, you will fritter your time away on conflicting priorities. People tend to neglect goal setting because it requires time and effort. What they fail to consider is that a little time and effort put in now saves an enormous amount of time, effort, and frustration in the future.

Goal setting is a powerful process for thinking about your ideal future and for motivating yourself to turn this vision of the future into reality. The process of setting goals helps you choose where you want to go in life. By knowing precisely what you want to achieve, you know where you have to concentrate your efforts. You will also quickly spot the distractions that would otherwise lure you off course. Additionally, properly set goals can be incredibly motivating, and as you get into the habit of setting and achieving goals, you will find that your self-confidence builds fast.

A useful way of making goals more powerful is to use the SMART mnemonic. Though there are plenty of variants, SMART usually stands for

- S: Specific
- M: Measurable
- A: Attainable
- R: Relevant
- T: Time-bound

For example, instead of saying "I will lose 30 pounds" it is more effective to say, "Starting January 1st, I will go on a diet and lose 30 pounds by Labor Day of this year."

- *Prioritization:* Prioritizing what needs to be done is especially important. Without prioritization, a person may work very hard yet not achieve the desired results because what he or she is working on is not of strategic importance. Most people have a to-do list or a top-ten list of some sort. The problem with many of these lists is that they are just a collection of things that need to get done. There is no rhyme or reason to the list and therefore the work they do is just as unstructured. So how do you work on to-do list tasks—top-down, bottom-up, easiest to hardest? To work efficiently, you need to work on the most important, highest value tasks. This way you will not get caught scrambling to get something critical done as the deadline approaches.

- *Managing Interruptions:* Having a plan and knowing how to prioritize it is one thing. The next issue is to know what to do to minimize the interruptions you face during your day. It is widely recognized that managers get very little uninterrupted time to work on their priority tasks. There are phone calls, information requests, questions from employees, and a whole host of events that crop up unexpectedly. Some do need to be dealt with immediately, but others need to be managed. An excellent tool that may be used is the urgent/important matrix. However, some jobs need you to be available for people when they need help—interruption is a natural and necessary part of life. Here, do what you sensibly can to minimize it, but make sure you do not scare people away from interrupting you when they should.

- *Procrastination:* "I'll get to it later" has led to the downfall of many a good employee. After too many "laters" the work piles up so high that any task seems insurmountable. Procrastination is as tempting as it is deadly. The best way to beat it is to recognize that you do indeed procrastinate. Then you need to figure out why. Perhaps you are afraid of failing or even success? Once you know why you procrastinate, you can plan to get out of the habit. Reward yourself for getting jobs done and remind yourself regularly of the horrible consequences of not doing those boring tasks.

- *Scheduling:* Much of time management comes down to effective scheduling of your time. When you know what your goals and priorities are, you then need to know how to go about creating a schedule that keeps you on track and protects you from stress.

  This means understanding the factors that affect the time you have available for work. You not only have to schedule priority tasks, you have to leave room for interruptions and contingency time for those unexpected events that otherwise wreak havoc with your schedule. By creating a robust schedule that reflects your priorities as well as supports your personal goals, you have a winning combination: one that will allow you to control your time and keep your life in balance.

Perhaps one of the things that we should concentrate on is the focal point of procrastination. If you have found yourself putting off important tasks over and over again, you are not alone. In fact, many people procrastinate to some degree, but some are so chronically affected by procrastination that it stops them from achieving things they are capable of and disrupts their careers. The key to controlling and ultimately combating this destructive habit is to recognize when you start procrastinating, understand why it happens (even to the best of us), and take active steps to better manage your time and outcomes.

In a nutshell, you procrastinate when you put off things that you should be focusing on right now, usually in favor of doing something that is more enjoyable or that you are more comfortable doing. We all must remember that procrastinators work as many hours in the day as other people (and often work longer hours), but they invest their time in the wrong tasks.

Sometimes this is simply because they do not understand the difference between urgent tasks and important tasks and jump straight into getting on with urgent tasks that are not actually important. They may feel that they are doing the right thing by reacting fast. Or they may not even think about their approach and simply be driven by the person whose demands are loudest. Either way, by doing this, they have little or no time left for the important tasks, despite the unpleasant outcomes this may bring.

Another common cause of procrastination is feeling overwhelmed by the task. You may not know where to begin. Or you may doubt that you have the skills or resources you think you need. So you seek comfort in doing tasks that you know you are capable of completing. Unfortunately, the big task is not going to go away—truly important tasks rarely do. Other causes of procrastination include but are not limited to the following:

- A fear of failure or success.
- A fear of perfectionism (I cannot do it perfectly now, so I would not attempt it).
- Waiting for the right mood or the right time.
- Not knowing what to do.
- Poor organizational skills, including underdeveloped decision-making skills.

Great time management means being effective as well as efficient (see Table 6.1). Managing time effectively and achieving the things that you want to achieve means spending your time on things that are important and not just urgent. To do this, and to minimize the stress of having too many tight deadlines, you need to distinguish clearly between what is urgent and what is important:

- *Important* activities have an outcome that leads to the achievement of your goals.

TABLE 6.1

Basic Form of an Action Priority Matrix

		Projects
Impact		
High	Easy accomplishments	Major projects
Low	Impromptu projects	Required but lengthy projects that nobody wants to get involved with
Effort	Low	High

- *Urgent* activities demand immediate attention and are usually associated with the achievement of someone else's goals or with an uncomfortable problem or situation that needs to be resolved. Urgent activities are often the ones we concentrate on. These are the "squeaky wheels that get the grease." They demand attention because the consequences of not dealing with them are immediate. These activities push us toward "firefighting" rather than being good planners.

SCHEDULING

People who deliberately undertake to schedule their time are not the ones who have decided to spend all their time on specific things at the expense of anything else. They usually have decided to efficiently use the time they have to spend a certain time on specific tasks and to desensitize themselves to the many distractions that commonly occur. What does this desensitizing involve? It means removing oneself from constant day-to-day, hour-to-hour decisions as to whether one will or will not spend the next hour on routine issues or will not go to a show on impulse and whether or not to use that time to get ready for next day's or week's tasks out of the way.

A workable time schedule can make help you make decisions, thus desensitizing you to momentary distractions. Usually a minimum time schedule is best. In other words, plan what you know is necessary and add to it later only if necessary. But plan as your first schedule one that you know you can keep and one that it is important to you to keep.

SUMMARY

In this chapter we focused on time management. Specifically, we discussed how time management plays a role in an individual's efforts toward improvement. This is somewhat of a different approach compared with project management, which is a project orientation method for improvement. In the next chapter we will address the concept of engagement and its importance to overall quality efforts.

SELECTED BIBLIOGRAPHY

Covey, S. R. (2004). *The 7 habits of highly effective people*. New York: Free Press.

Fiori, N. (2007). *The now habit: A strategic program for overcoming procrastination and enjoying guilt-free play*, revised ed. New York: Tarcher.

Griessman, B. E. (1994). *Time tactics of very successful people*. New York: McGraw Hill.

Lakein, A. (1989). *How to get control of your time and your life*. New York: Signet.

MacKenzie, A. (1997). *The time trap*, 3rd ed. New York: AMACOM.

7

Engagement

In Chapter 3 and Chapter 6 we focused our discussion on project management and time management, respectively. Our purpose was to make the reader cognizant that both project and individual management are important for a completed task. In this chapter we move even more toward this completion goal with a discussion on engagement. Here we concentrate on how the leader of the organization communicates to the staff and other people involved with a project so that everyone is supporting the vision of the leader. In essence, we concentrate on how to maximize energy so that a task is completed with the least resistance.

As we have said several times, success depends on the execution and results of projects. So, the question for the future is how we get to be good executioners with positive results. A simple answer is through engagement of the entire organization toward specific goals that have been defined through strategy, vision, and so on. *Engagement* in this case is interpreted as "managing energy to maximize capacity."

In Chapter 3 we addressed project management and PRINCE2. We also made reference to the Project Management Institute (PMI), which has created a detailed outline of the project management body of knowledge (PMBOK) that outlines the various inputs, tools, techniques, and outputs of projects from initiating to closing. The knowledge areas include integration, scope, time, cost, quality, human resources, communications, risk, and procurement management.

Though PMBOK and PRINCE2 offer an excellent rationale guide to project management, they are somewhat short on the assurance of what you do to ensure full engagement when people have so many different agendas and so many tasks on their overflowing work "plates." The answer is of course to provide appropriate and applicable engagement policies as well as understand the issues of communication, delegation, and change management.

Engagement in terms of project management refers to each stakeholder's investment of energy, skill, ability, effort, and eagerness in a defined project. This includes involvement and commitment yet goes beyond, to include various actions such as:

- Attends to task details
- Builds strong connections and relationships
- Balances support and challenge with inquiry and advocacy
- Speaks highly about the project to others
- Commits to project completion
- Involves self in special subprojects
- Communicates willingly and effectively with others
- Demonstrates personal/professional development
- Initiates problem solving and/or conflict resolution
- Innovates effective processes and procedures

The reader should notice that we mention *energy* and not time as the more precious—*time* is always finite and the same for every one (24 hours per day), whereas energy is always dynamic and depends on four dimensions: (a) body, (b) emotions, (c) mind, and (d) spirit. Therefore, it is this combination of the four dimensions that makes the real meaning of capacity and how it brings skill and talent into life for an organization to be successful. *Capacity* here is viewed as the project or vision of management for the organization. It may be focused on a variety of issues, concerns, or projects in one or several departments of the organization, such as quality, financial, production, customer service, and so on. In addition to engagement, we must be very careful about communication and delegation (see Appendix E on the companion CD-ROM for opening the doors to communication).

To understand the energy in all of us we must understand the emotions that guide us throughout the day in all of our endeavors. These emotions, from our experience and many psychological, psychiatric, and other sources, may be summarized as:

- Survival for all
  - Enthusiasm
  - Interest
  - Conservatism
  - Boredom

- Survival for self
  - Antagonism
  - Anger
  - Concealed hostility
  - Fear
- Survival for no one
  - Sympathy
  - Propitiation
  - Grief
  - Death

The above summary does not take into account work, exhilaration, and enlightenment, because they are not emotions. Whereas these are rough observations and generalizations, the problem is that these are forever changing. If, however, we use emotions as a model, a consistent pattern emerges that is highly usable (but very short sighted) in practice. The model becomes that of fear and intimidation. When that happens, the result is always short-term gains, overruns due to inefficiency, and low morale of all participants in the specific project.

Stakeholders are anyone who has an interest in the project, including project leader, team members, upper management, customers, resource manager, etc. Project stakeholders are individuals and organizations that are actively involved in the project or whose interests may be affected as a result of project execution or project completion. They may also exert influence over the project's objectives and outcomes. The project management team must identify the stakeholders, determine their requirements and expectations, and, to the extent possible, manage their influence in relation to the requirements to ensure a successful project.

I believe this is where active project engagement can play an assessment and ongoing role in contributing to the success of the project. Here are some initial questions involving project engagement:

- If you are a project manager, how much attention do you pay to project engagement?
- How do you engage yourself in projects?
- How do you assess engagement levels of the various stakeholders?
- What tools and techniques can you use to foster fuller levels of engagement?

- How do you customize your project engagement efforts for various stakeholders?
- How do you keep your team engaged in the project?
- What factors may create project disengagement?
- How do you address project disengagement?

When we understand the engagement process as an employee engagement for the entire organization, we enrich everyone in the workplace. Authentic employee engagement must function for the benefit of all: employees, leaders, organizations, and customers. That means that engagement in a specific project is not taking more discretionary effort from employees; rather, employees have higher levels of satisfaction and contribution. Leaders are connected with employees and engaged themselves. The organization is functioning the way it should, and customers are receiving the service they deserve. Ultimately everyone is getting results that matter to them. When this occurs, employee engagement transforms into a more powerful force—workplace engagement.

COMMUNICATION

To effectively communicate, we must realize that we are all different in the way in which we perceive the world and use this understanding as a guide to our communication with others. When we hear certain people speak beautifully, many of us ask ourselves, "Oh how great it would have been if I could speak like that!" In some other interaction, we may feel that we should somehow opt out because of the boring or irritating conversation (Rizwan 2007).

Where does the difference between the two lie? It may be true that certain people may be gifted with a flair for speaking well. They may be, though not deliberately, adopting styles that make speech useful and interesting. We suggest that good communicators are not "square." They are, in fact, adding a dimension of enthusiasm, care, and so on. This is shown in Figure 7.1. Let us look at the styles in some depth.

Unlike writing, speaking is an integral part of your everyday life. Whether you are an unlettered villager or an erudite scholar, a child or a grandfather, you have to speak. This may make us feel that speaking well is natural; you take it for granted. This assumption is not true. If you want

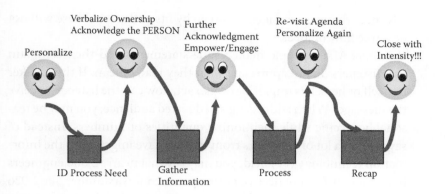

FIGURE 7.1
Don't be just a square!

to speak well, you should take care of several things. Perhaps one of the most important things that one should be aware of is that there is no uniform style that fits all occasions. For example, the language you use while bargaining in a fish market will be totally different from your language during a professional presentation or a group discussion for job recruitment. Even so, there are common attributes that make speaking effective, including the following:

- *Clarity:* The most essential characteristic of good speech is clarity. Clarity of expression can come only from clarity of thought. A cluttered mind cannot come up with clear ideas. Your ideas have to be clear. Some say that you have to speak with the brain and the mouth, if you are to speak well. In other words, think before you speak. You have to use simple words and simple expressions if others are to easily understand what you say.

   Those with a very rich vocabulary or jargon may be tempted to flaunt it. If listeners are unsure of the meaning of the words you utter, you lose clarity and challenge comprehension. Also, the use of professional jargon before laymen should be avoided if you want to convey a message effectively. Professionals like doctors or engineers (including quality professionals of all levels) should not use technical terminology while speaking to those outside their profession. The use of nonstandard abbreviations will naturally reduce the clarity of what you say.

- *Logical organization:* The change from one point to another should be smooth and sensible. The sequence of ideas should be planned in advance wherever possible. If you go by fits and starts, there will not be a free flow of ideas.

- *Precision:* Many people make vague statements and then complain that listeners do not appreciate what they really mean. If the speaker himself or herself is imprecise or inexact, how can the listeners receive the message? When addressing an educated audience, you may be reasonably specific while mentioning quantities or numbers. Instead of saying that a lot of engineers from the city have migrated to the information technology (IT) field, you may say that nearly 2,000 engineers have done so. It is not necessary that you mention a number like 2,026 engineers, unless a detailed statistical exercise is involved.

- *Audience:* Analyze the audience before planning the nature and style of speech.

- *Catch phrases:* If you can use some attractive phrases that would effectively draw their attention to your subject, the rest of your job becomes easy. The phrases need not be pedantic but may be made up of ordinary words that when grouped render an extraordinary feeling.

   Remember how Gandhi wisely used "Quiet India" to electrify the people during the freedom struggle. *Citius, Altius, Fortius,* meaning "Swifter, Higher, Stronger," is the motto of the Olympics. "Ornaments for Armaments" was coined to promote gold donations to the National Defense Fund in India during the Chinese aggression in 1962. A very late political message used in the 2008 presidential political campaign of the United States of America was the slogan "Change—Yes we can." Even though quite nebulous in content, it brought fantastic results. There are, of course, many other interesting examples.

- *Courtesy:* Respect the listener while you speak. None ever lost anything through courtesy or politeness, whereas arrogance has led to many reverses. Listeners generally do not like speakers who do not show empathy. The best way to show empathy is to put yourself in the other person's shoes. The message of courtesy can easily be transmitted through nonverbal communication as well.

- *Unwise generalization:* You must have heard unconvincing political speeches, with expressions such as "the people of this country are with us" and "no sensible person will agree," and so on. The people who are with them may be just a fraction of our population. There may be several sensible people who agree with the opposite views. Still, they make such generalizations with the hope of carrying others with them. You have to avoid such generalization if you want to be convincing and effective in speaking.

- *Sincerity:* Your words should ring with sincerity. Your views should emerge from your heart. Try to be natural; do not try to be another person. Look with confidence into the eyes of the listener.
- *Goal:* Keep your goal in mind during all stages of speaking. Speak with confidence to attain the goal.
- *Illustrations:* Use examples and illustrations (whenever a complex idea has to be presented). Further, go from the simple to the complex to ensure comprehension. Relate personal experience of relevance.
- *Dress:* When you address a gathering, you should dress for the occasion. How well groomed you look sends a definite message. You are indirectly respecting the audience by dressing properly. It is professional dress, but not expensive attire or gaudy clothing, that impresses an audience. Not only your dress but your overall appearance has to be elegant and pleasing.

Posture and personal cleanliness should not be ignored. Your looks and movements should radiate energy and enthusiasm. The key factor is that the audience should feel that you deserve sympathetic listening; if you pass this stage successfully, half the battle is won.

- *Style of delivery:* Controlling your voice to suit the occasion is important. Various parts of your body such as mouth, teeth, tongue, throat, vocal cords, lungs, jaws, muscles in the upper part of the trunk, larynx, and diaphragm are involved in the speech process. Proper manipulation of these parts results in the control of various qualities, like pitch, volume, tone, and tenor of speech.

It is not that you manipulate these parts deliberately, but when you try to control the voice, appropriate changes take place automatically as required. Never talk too fast or you may lose clarity. Never be tiresomely slow; you will test the patience of listeners. Make use of pauses for emphasis. Never raise your voice too high. Never force the listeners to strain their ears by keeping the pitch too low. Speak distinctly to retain clarity. Keep in mind the energy level of the listeners. Sprinkle your speech with elegant humor, if it suits the occasion.

- *Pronunciation:* How you pronounce words is also important. Accent and diction have to be kept in mind while speaking. If necessary, you may record your own voice and delete and make corrections based on feedback from your own recordings.

- *Mannerisms:* Identify and avoid mannerisms, both verbal and non-verbal, so that they do not distract the listeners.
- *Feeling:* Put feeling into your voice and make your speech expressive. You should never sound bland or insipid. You have to be enthusiastic if you want listeners to fix their attention on you.
- *Alertness:* You cannot afford to be casual if you want your speech to be effective. There has to be a great amount of alertness in your approach. If you keep all of these points in the back of your mind when you speak and try to follow the best possible styles to suit a given occasion, you are sure to succeed in communicating effectively.

The mantras these days are about skills, workplace training, the learning organization, and lifelong learning. But it seems that this applies to everyone except senior executives. Though nine out of ten first-line managers received training in 2008, only six out of ten senior executives were trained with them, according to a study by Boston-based consulting and training firm Novations Group (2008).

The problem is there are many first-line managers who might be brilliant technicians. They get promoted to a senior role and suddenly they are out on their own, in a world where resources are tight, pressures intense, and the politics toxic. It is the kind of role where support and training are important.

The Novations Group study suggests that we underestimate the amount of training senior executives receive (2008). And it tells us that senior managers are probably ambivalent about training. They would say they have too many other things to do. This issue was raised by Ackoff, Addison, and Bibb (2007). They set out 81 laws (presented as *f*-LAWS) that show why all the conventional thinking about management is a crock. One of those *f*-LAWS is "The higher their rank, the less managers perceive a need for continuing education, but the greater their need for it (Ackoff, Addison, & Bibb 2007, p. 98)." That is, unless you hold the training at an exotic resort, preferably with a golf course, restrict the learning to only a few hours a day, and let them bring their wives or partners (*f*-LAWS nd).

If you are in a senior role, how much management training do you get? Could you do with more? Do you have time? For those lower down the chain, do your senior managers need more training?

Elegant practices of communication can, of course, be learned. However, one of the items that is generally overlooked is the communication process of multinationals in this globalized environment. I happened to be in

China during the 2007 Chinese New Year. The newspapers and the general media were in full discussion about the difference between Western and Chinese learning styles and its implications for Western managers working in China. The main difference is that Western learning culture is based on a questioning approach, whereas the Chinese learning culture is based on model behavior. As a result, Western managers find themselves challenged by silent meeting rooms, lack of initiative from their local staff, or difficulty in getting feedback when necessary.

Karlson reports that in most Western countries, leaders are used to working in an environment where the employees are highly participative in meetings and contribute their opinions and experiences to the decision-making process (2009). Their education is based on assuming a questioning approach to learning and not readily accepting facts given to them. This makes a Western employee feel more comfortable in arguing for his beliefs and sharing his opinions in meetings with superiors. Furthermore, a manager in the West is not necessarily assumed to be an expert in his field or industry. On the contrary, he is supposed to utilize the knowledge base of his entire team in decisions, projects, and even daily interaction and work. Western managers are therefore used to leading highly interactive teams and rely on two-way communication. When a team achieves strong synergy, the leader can and will step down, letting the team become highly innovative and self-managing. In doing so, the manager may focus more on strategic decisions and leadership tasks. Successful innovative Western companies have a flat organization, with managers leading teams with a democratic approach: coaching and supporting them.

Chinese people are taught to learn by listening and modeling their behavior on that of superiors and teachers. The general assumption is that a questioning approach, especially toward superiors, is disrespectful. It is professional in China to respect and fully support your superior, listen to your instructions, and follow the leader's example. A Chinese manager is not expected to rely on his team for knowledge. In fact, he is supposed to provide all of the answers and decisions. The staff is supposed to follow orders and carry them out swiftly and appropriately. The manager is expected to be autocratic, and in meetings and interactions mainly one-way communication is carried out. In the Chinese environment, the success of the company is therefore highly dependent on the skills and knowledge of the leaders. Great Chinese companies are frequently led by autocratic managers with great technical knowledge in their field and industry, enabling them to lead subordinates by setting clear outcomes (Karlson 2009).

Western managers arriving in China commonly use a democratic and collaborative leadership style. This often leads to great confusion among Chinese employees. Instead of having clear tasks and outcomes given by their new boss, they are included in meetings and asked to contribute to decisions that they never had the authority for before. Without detailed instructions to follow and no clear orders from their Western leaders, Chinese employees feel uncomfortable and insecure. The local employee asks himself how he is expected to do a good job if even his leader does not know what needs to be done. At the other end of the table, the Western manager feels frustrated with the team, because the members lack initiative and do not share opinions or contribute to the decision-making process.

Can Western managers succeed in achieving two-way communication and leading with a democratic leadership style in China? The answer is yes. We have seen many successful companies who have achieved strong collaboration and high levels of synergy in China. These Western leaders started out with a direct leadership style. Slowly but surely they started coaching their employees in taking small steps and assuming larger responsibilities. These companies have broken the communication barrier and managed to introduce two-way communication in their workforce. In many cases, these Chinese company divisions have outperformed all divisions in other countries.

DELEGATION

Another overlooked issue of communication is delegation. Delegation is crucial to success. Spread the work around, and more gets done in the long run. Delegate especially well and you can build a network of effective, highly motivated people who can bring you great success. But what is a good or even a superb delegator? Ensman (2005) has identified ten mistakes that one may avoid and in the process become a great delegator.

> Mistake #1: No plan. Master delegators take the time to plan the work with subordinates, so they can "work the plan" without extensive oversight. Neglect a good plan, and you will be tempted to constantly check in, criticize, or micromanage.

The solution: Ask your subordinate to develop a step-by-step plan—complete with timeline and reporting responsibilities—that you can review and agree on.

Mistake #2: No relationship. Do you have a quality relationship with your subordinates? If you have not built enough rapport and trust to work together without standing side by side, you will end up with problems.

The solution: Spend some old-fashioned quality time with your subordinates. Understand their skills, strengths, and weaknesses—and get to a point where you are comfortable discussing them.

Mistake #3: No candor. It is easy to sing the praises of delegation. But it is also easy to forget to give fast, genuine feedback—good and bad—to your delegates.

The solution: Each time you meet with a subordinate, identify any positive or negative reactions to his or her performance and discuss them.

Mistake #4: No communication. Delegating a task is never enough. You can easily forget to review the parameters of the assignment when it is made and forget the importance of check-ins while it is being carried out.

The solution: A review session (which can be part of a routine supervisory meeting or staff meeting) at the outset and opportunities for regular discussion later.

Mistake #5: No time. Delegated work may involve other people's schedules or assignments. If you fail to take these into account when delegating a task, your subordinates may find themselves unable to meet your deadlines because they do not have enough time to meet the expectations and schedules of others.

The solution: At the outset, identify all individuals—colleagues, vendors, customers, members—who must be involved in the task. Develop a mutually agreed-upon timeline for completion of activities. If the task is significant, get buy-in from all involved.

Mistake #6: No flexibility. How many times have you delegated responsibility only to hear later on that things did not work out because of changed circumstances? It happens!

The solution: Give your subordinate the discretion to make decisions when confronted with information or problems that you did not originally anticipate.

Mistake #7: No accountability. Once you give someone responsibility for a task, how do you ensure that it is properly completed? If you do not put a mechanism for accountability in place, you will be increasing the chance of a poor outcome.

The solution: Regular reports, exception notifications, and special meetings.

Mistake #8: No clarity. When you delegate, do you explain exactly what the tasks involve or what outcomes you desire? No? Then you might end up in the wrong place.

The solution: Give your subordinate a "charter" when delegating a task—a clear statement of responsibility repeated several times (with your subordinate affirming his or her understanding) or a formal written assignment if the task is large.

Mistake #9: No confidence. If you ask a designated subordinate every few hours how a task is going or repeatedly criticize during check-ins, you will send a clear message to your subordinate: "I don't think you can do this task properly." And guess what? If your actions suggest this, he or she probably will not be able to do it.

The solution: Look for opportunities to praise your subordinate during the process—the occasional compliment, a thank-you note, a word of encouragement during a staff meeting. And assume the role of coach, encouraging shared problem solving.

Mistake #10: No consistency. How many times have you delegated a task only to signal changes in your expectations with regularity? Shift your position, and you will slow the completion of the task—and increase the chance of an unsatisfactory outcome.

The solution: Agree on the outcome, the reporting mechanisms, and the amount of authority your subordinate has in advance. Change your expectations only in the event of a serious problem.

RISK AND PEOPLE

When we talk about *globalization*, by definition we talk about multinational risk in all aspects of the organization. That means for the home organization but also for the host country. As we mentioned in Chapter 3, managing risk is a challenge at the best of times. It is even harder in the context of a rapidly changing marketplace where the risk profile is always evolving. However, the advantage of getting it right is quite significant. Businesses with a systematic and controlled approach to risk can confidently make decisions regarding both the threats and opportunities that any risk may present because these companies are better prepared for the unexpected and the unintended.

The key connections between globalization and risk are the issues of communication, culture, and local norms. How we communicate, how we understand the culture, and what the typical behaviors and expectations are in the global market are of importance.

For example: Western executives look to countries such as China, India, Thailand, Indonesia, Mexico, Vietnam, Cambodia, Brazil, and Russia as large and growing markets. But although they offer great potential rewards, emerging markets are also rife with potential dangers that are rare in more developed countries. Risks endemic to emerging markets include rapid and dramatic fluctuations in exchange and interest rates, political turmoil, murky intellectual property regimes, unpredictable flip-flops in industrial policy, exposure to volatile commodity prices, and uncertain access to capital.

A classic example of this volatility was Brazil in the 1990s. The country had to endure hyperinflation (exceeding an annual rate of 2,000 percent in some years), followed by an abrupt price stabilization that ended overnight the practice of passing cost inefficiencies through to customers. Bills were paid in four separate currencies among dramatic swings in exchange and interest rates. The reader should notice that we do not even mention the political turmoil and industrial policy changes from President Fernando Collor de Mello, who eliminated tariffs and domestic content requirements in the automobile components sector, causing 19 of the country's top 20 suppliers to file for bankruptcy or sell to multinationals. In 1992 the country had to deal with energy rationing, technological competitiveness, customer uncertainty, and recession.

Sull (2005), in his famous research on risk in the globalization market, identified the approach to risk management as the most consistent difference between the successful and less successful companies. The lessons that Sull presented from these emerging market leaders offer insights into managing risk in any unpredictable market and are particularly relevant for managers expanding into emerging markets from more predictable competitive domains.

Western managers generally view risk management in purely negative terms as a set of mechanisms to minimize the probability and/or cost of undesired outcomes. However, one finding from Sull's (2005) research is that executives in unpredictable markets typically view risk more broadly—based on long experience, they recognize that the volatility inherent in their domestic markets entails both potential hazards and opportunities.

A key finding in Sull's (2005) research is that successful managers in emerging markets rely on the same five mechanisms—anticipating contingencies, sharing risk, keeping a reserve, diversifying, and picking the right people—to both help minimize the probability and cost of undesirable outcomes and increase the probability of seizing potential opportunities. This reliance on the same mechanisms to capture upside and manage downside risk strikes many Western managers as strange.

- *Anticipate:* Managers in emerging markets must advance into an unpredictable future, with multiple variables influencing their ability to create, capture, and sustain value. But though they cannot predict the long-term future with great precision, executives can take steps to anticipate emerging threats and opportunities.

  Companies in emerging markets explicitly scan their environment by gathering data from multiple sources, conducting studies, and discussing scenarios. In some cases they take part in governmental meetings. This is especially true in China, where you find executives who belong to the Communist Party have access to inside information. Western executives often put too much faith in tools such as scenario planning or simulations. Emerging market executives harbor more realistic expectations for anticipation, generally limiting their exploration to events that are likely, even if their timing or precise implications are unclear.

  A classic example of this anticipation is the Guangdong's Galanz Company, today the world's leading microwave oven producer. In 1993, it was a goose down producer looking for new markets to offset declining sales in its core. To explore potential demand for microwave ovens, Galanz executives negotiated a limited technology license with Toshiba, produced a trial run of 10,000 ovens, and sent senior executives to a Shanghai department store to personally pitch the new appliances to skeptical consumers. Galanz rapidly scaled nationally after this initial experiment supported management's hunch that the microwave oven market was poised to take off.

- *Share:* Executives generally consider risk sharing in narrow terms of transferring a potential downside to another party; for example, by using insurance contracts or currency hedges. However, it can also enable companies to seize opportunities.

Shifting noncritical activities to partners that specialize in them can also enable companies to focus on core activities, thereby increasing their odds of successfully seizing an opportunity. A classic example of this is the Brazilian company Embraer. Its suppliers bore two-thirds of the total development costs for the new jet.

• *Diversify:* Diversification protects companies against unforeseen events that threaten one of their core businesses, but it also exposes them to a broader range of opportunities. For example, Embraer produced both military and commercial aircraft, and Galanz added air conditioners to microwaves. Though diversification opens up opportunities, it also increases organizational complexity and can spread constrained resources (including management attention) too thinly. In turn, complexity and diffuse resources may limit a company's ability to compete effectively against more focused rivals.

Diversification in the global market is risky; however, according to Sull (2005), there are several practical steps that managers can take to reap at least some of the rewards of diversification without spreading themselves too thinly. First, they can follow the rule of two by pursuing two markets simultaneously, which provides some diversification while minimizing the problems associated with complexity and resource diffusion. Managers can also diversify geographically as well as by a set of specialities while retaining a tight industry focus.

• *Reserve:* In contrast to specialized resources, cash is fungible—it can be deployed against most contingencies. Fungibility is a particularly valuable attribute in markets where it is difficult to predict which resources will confer future sustainable competitive advantage. Cash functions like troops that a general keeps in reserve until seeing an opportunity to win a battle or defend important ground.

On the upside, cash allows companies to invest in whichever specialized resources are required to seize the moment. Building a reserve of cash is particularly important in contexts where the cost and availability of capital is subject to wide swings. In developed countries, good opportunities typically find funding and sound companies facing temporary setbacks can raise the cash required to tide them over. However, this efficient market view bears little resemblance to the reality of fundraising in emerging markets. In these countries, funds are least likely to be available precisely when a company hits a setback. A good example of this is the Russian ruble crisis, which deprived even healthy companies of required funding. Even if financing is available, the fundraising process demands time and management attention, and capital raised under duress generally comes at a high price in terms of valuation, preferences, and control.

Though a large reserve provides a powerful hedge against uncertainty, there are obvious disadvantages to stockpiling cash. Managers may become sloppy and squander funds on fancy buildings or unsound acquisitions. Building a war chest can also diminish a company's autonomy, because capital providers generally demand control in return for their money.

- *People*: Employees can play a vital role in a company's ability to manage risk. The obvious example occurs when a company faces a clearly defined opportunity or threat and hires someone with experience in that area; for example, Sina.com, China's leading Internet portal, which faced a meltdown after the Internet bubble burst in 2001 and replaced its CEO with a venture capitalist experienced in restructuring and dealing with investors.

To succeed consistently at dealing with unexpected contingencies, companies in volatile markets require a cadre of managers who can quickly and effectively deal with new challenges, rather than doing the same thing over and over again. Managers who excel at these activities tend to share a few characteristics: they take initiative; enjoy autonomy; have excellent communication as well as delegation skills; and have a diverse history of assignments in terms of functions, regions, and tasks. An example of this is the Haier Corporation. Haier rose in 20 years from a bankrupt collective enterprise to China's largest appliance maker. Over those years, Haier faced a series of threats and opportunities. The company thrived in large part because it had general managers who could switch between building a new industrial park and integrating an acquisition or opening a foreign subsidiary without missing a beat.

This particular company is effective at producing general managers with a broad view of interrelationships among the various parts of Haier and who are expert in quickly assessing a new situation and executing objectives. Their effectiveness depends on Saturday morning training sessions; working in diverse teams to brainstorm ways to achieve their individual business objectives; the opportunity to practice their learning during the week; and rotation on average every 3 years between different products, regions, and functions.

MANAGING CHANGE

As we mentioned in the Preface, change in any environment is necessary. However, how this change happens is of importance and needs to be understood by all concerned. In any change there is a resistance for the status quo. However, resistance is directly tied to the level of disruption (introduction of the change).

That resistance may be of the organization or the people. In either case, something has been recognized as inefficient, old, costly, ineffective, not trustworthy, low reward, fear of the unknown, and so on. These obstacles, no matter how many or what their magnitude, follow a certain pain path that reaches the goal of the proposed change. The path follows a normal distribution curve with the x-axis as time and the y-axis for the effort against the change.

On the left of the curve one notices the beginning of the pain as the immobilization (introduction of the change) occurs. It is followed by denial (people refuse to accept the change); this is followed by anger, which is the highest resistance level. At this point, individuals are very uncomfortable with the change and they begin to question whether or not the change will solve anything or whether their job will be in peril. This is followed by bargaining (negotiating the change) and then depression (feeling uncertain about the proposed change). Once the realization has set in that the change is happening, the exploration (finding and recommending alternatives) stage begins and finally the acceptance stage occurs. In this stage everyone realizes that the change is for real and they have to accept it.

Once the pain path is understood, then the management process for the change begins. It is important to understand the pain path because in the process of change there will be successes and failures. They are to be expected. If you know what it takes to introduce the change, you can plan accordingly and minimize frustration due to failures and/or setbacks that occur.

Preparation is the essence of managing any change. Specifically, one must recognize that the drivers for change may be external or internal to the organization. In other words, it is imperative to identify the change and determine the key elements for that change as early as possible. It is essential to analyze the current state (where we are), define the desired state (where we want to be), and assess the change itself by asking what is the current state.

Specifically, for the current state typical questions that may be asked include the following:

- Why is there a need for this change?
  - What would happen if we did not change?
  - What change forces are internal? External?
  - How much time do we have to change?
  - What if our attempt to change fails?
  - Did it hurt or help us not to do this earlier?
  - Why didn't we do this earlier?
- What is the current situation?
  - What has to change?
  - What is going wrong?
  - What needs are not being met?
  - What resources are unused or underutilized?
  - What opportunities are not being exploited?
  - What challenges are not being met?
  - What do competitors do better?
- What will get in the way of the change?
  - How did things get this way?
  - How long have things been this way?
  - What keeps things the way they are?
- How powerful is the current state?
  - Are people comfortable with the way things are?
  - Are people skilled at doing things the current way?
  - Are people satisfied with the results of their work?
  - How much energy do people use to get the work done?
  - Do certain jobs and titles carry more prestige than others?
  - Does the way work gets done make sense to those doing it?

For the desired state, management must be concerned with the future. Because this desired state of change is future oriented, we must recognize that it can be an evolutionary change or a breakthrough change. In either case, both have to be designed and planned accordingly with short- and long-term planning. For example, in the case of a merger or acquisition, management should be concerned with whether or not they should follow the path of

- Coexistence (separate but supportive cultures)
- Assimilation (dominant culture prevails through attraction or coercion)
- Transformation (development of a new culture)
- Rejection (separate and hostile cultures)

Culture as an element of change is how we act, what we believe, and the rules we follow. So, when management is reviewing the culture of the organization, they must be aware of the unwanted behaviors, beliefs, and rules. On the other hand, they must also have a vision and an understanding of what they want in the way of desired behaviors, beliefs, and rules. (*Rules* are the formal policies and procedures as well as the unwritten rules that we follow in the course of performing a task.) The reason for this knowledge is that no change exists in isolation. In fact, changes are more often than not ongoing and usually moving targets in areas such as operating procedures, information technology, organizational structures, product and services, work processes, and so on. As such, management must be prepared for some kind of a ripple effect. This preparation may begin with the following questions in regards to integrating the change(s):

- Are these all the major change initiatives?
- Are these desired state outcomes agreed to by all key sponsors?
- Is there a conflict among these change initiatives?
- Is there a logical sequence of implementation?
- Are there common resource requirements?
- Could everyone in the organization identify these changes and define them?
- Could everyone in the organization accurately define themselves in the desired state?

The reason for these questions is that any change in any organization, whether pending or current, is difficult. In fact, any change is not safe, it is not guaranteed for success, it is expensive, it is sad, it is exciting, it is stressful, it is not the old way or the new way, it is rewarding, it is challenging. Therefore, the gauge of change must be on the balance of the data

TABLE 7.1

The Gauge of Change

What is it about the present that makes the targets secure or safe?		What will cause people problems in the transition?		What is threatening about the future state?
Present situation (Today)	+	Change	–	Future = +/–
Present situation (tomorrow)	+	Change	0	Future = +
The Present: Why do people stay here? What are the Formal and informal rewards here? What logical things hold them here? What still works well here? What does not work as well as it did in the past? What will happen if we continue to stay here?		**The Change:** What would make people feel insecure or scared? What would drain company resources? What would make people feel sad? What would make people excited? What would make people feel stressed?		**The Future:** Why would people want to be here? Why would people not want to be here?

gathered and analyzed based on risk and benefit analysis. At least in this stage it could take the form shown in Table 7.1.

Typical questions for the desired state are the following:

- What is the change?
- How does the change fit into the overall business strategy?
- What will the desired state look like?
- What elements of the desired state (if any) are negotiable during implementation?
- How stable will the desired state be?
  - What will remain constant?
  - What will be subject to ongoing change?
- How will the organization know when it has reached the desired state?
- How long will it take to reach the desired state?

Once that investigation is completed, we move into the next phase, which is preparation for the change. In other words, we must identify and evaluate the key stakeholders. Generally they are the sponsors, change agents, and targets.

The sponsors must understand their responsibility and target performance as it relates to change. For example:

- They must understand the change. That means:
  - Define the future as much as possible in both numerical and behavioral terms.
  - Build consensus for the desired state down through the organizational structure.
  - Identify and communicate what is firm and nonnegotiable and what is open to refinement and adjustment.
  - Show pictorially and operationally how this change fits into the framework of a larger change and/or how the change integrates with all other changes going on in the company.
  - Define the changes required of management to achieve the desired state.
  - Show understanding of the impact of the current culture and the history of change in the company on this change effort.
  - Commit to dealing with the degree and type of resistance that will surface regarding this change.
- They must understand how to manage the change resources. That means:
  - Allocate the dollars required for this change or explain why dollars are not available. If the dollars are not available, lower future expectations, provide alternative resources, or prepare for a less successful change.
  - Allocate time required for this change or explain why time is not available. If the needed time is not available, lower future expectations, provide alternative resources, or prepare for a less successful change.
  - Identify and empower effective change agents.
  - Make sure that the required change management systems are available:
    - Communication system
    - Learning system
    - Reward and reinforcement system
  - Identify the amount and type of loss tolerable in the change:
    - Productivity
    - Customer support

  - Profitability
  - Stress levels
- They must know how to deal with people. That means:
  - Exhibit empathy and understanding of the difficulties of change.
  - Communicate the desired state from the target's perspective, using the target's language.
  - Establish multiple listening opportunities to allow people to express resistance and fear.
  - Let people see that the sponsors are making changes and having difficulties.
  - Take risks and encourage others to take risks.
  - Provide a constant beacon of support and encouragement.
  - Coach other sponsors, change agents, and targets to develop their skills in managing change.

Furthermore, the sponsor must be very clear of their role. That means that they must understand what *sponsor* means and what their responsibility and competencies are all about. Without this understanding, successful change implementation will never occur.

Therefore, sponsors, by definition, have the authority, resources, and accountability to call for and support change. This translates into a series of responsibilities, such as:

- Demonstrate an understanding of the personal and organizational impact of a change.
- Understand how the many different change efforts in the organization affect one another and the overall mission.
- Secure commitment to the change from other sponsors: peers and superiors.
- Recognize and commit to changing themselves to support the change.
- Clearly articulate why the change is needed and their level of dissatisfaction with the current state.
- Keep the organization focused on the change.
- Predetermine how much disruption the organization can handle in the change.
- Support the change agents.
  - Hold them accountable.
  - Be available for them.
  - Let targets see the sponsor support.

- Hold the sponsor cascade accountable for implementing the change.

On the other hand, when we refer to *competencies*, that means:

- Understand and use the target's language and perspective.
- Solicit and respond to feedback.
- Develop good working relationships and trust at all levels.
- Demonstrate basic leadership skills.

Sponsors cannot delegate their duties to change agents. Sponsors are the first targets and sponsors are targets first!

An example of a simple evaluation for a typical sponsor may be the following, which consists of several statements. Obviously, these statements are not exhaustive ones but rather common and simple ones. They may be presented in a Likert-scale format (1–6, where 1 = *strongly agree* and 6 = *strongly disagree*).

- Has a clear vision of the desired state.
- Understands what it will take to achieve the desired state.
- Understands the external drivers of change and the cost of not changing.
- Is open and flexible regarding the desired state.
- Is open and flexible regarding the path to the desired state.
- Communicates personal commitment to the desired state.
- Communicates from the perspective of the target audience.
- Keeps self and the organization constantly focused on the desired state.
- Encourages people to express concerns and fears and to seek additional information.
- Recognizes and acknowledges the personal and organizational changes required throughout the organization.
- Is willing to provide the resources to achieve the desired state.
- Will align rewards/reinforcement to support the desired state.
- Requires feedback on measurable/observable increments toward the desired state.

The focus of this assessment is to make sure that the sponsor is the appropriate individual for the job and to make sure that the sponsor is communicating effectively throughout the organization. Typical items that may be communicated are the following:

- Red flags
- Help memos
- Panic buttons
- Code blue (emergencies)
- Opportunity sessions
- Note to boss's boss
- Other

The change agents must understand their responsibility and target performance as it relates to change. For example:

- They must understand the change. That means:
  - Identify and measure the impact of the change on the people, structures, processes, and culture of the company.
  - Determine the level and type of sponsorship required.
  - Find the appropriate sponsors and transfer ownership of the change to the right sponsors.
  - Figure out the primary and secondary sources of resistance by the target population.
- They must understand how to manage the change resources. That means:
  - Systematically apply the principles of effective change management:
    - Communication
    - Learning
    - Rewards and reinforcements
  - Integrate multiple changes into a common plan and apply sound project management principles to all of the changes required.
- They must know how to deal with people. That means:
  - Listen, listen, listen!
  - Translate the messages to targets and sponsors in language that is meaningful to them.
  - Build team strength at the project level, in the senior management group, among target groups.
  - Develop effective coaching and counseling techniques.
  - Build organization and structure into the change wherever and whenever possible.

- Use a process such as fishbone diagramming to identify, monitor, and track changes in the change.
- Use that process to keep the organization focused on changes in the change while maintaining regular company business.
- Know when to solve problems and how to do it effectively.
- Know when and what to delegate.

Furthermore, the change agent must be very clear of his or her role. That means that they must understand what *change agent* means and what their responsibility and competencies are all about. Without this understanding, successful change implementation will never occur.

Therefore, change agents, by definition, facilitate change. They are process owners, technical experts, and staff support who are competent, influential, and have a clear understanding of the business and its future direction. This translates into a series of responsibilities such as the following:

- Validate that the change is defined and well sponsored.
- Demonstrate a deep understanding of the personal, organizational, technical, customer, and supplier dynamics of the change.
- Build effective sponsorship.
  - Remember that sponsors themselves must practice what they preach.
  - Educate sponsors.
  - Balance the ownership of accountability between themselves and the sponsors.
- Assess the critical change variables.
  - Identify sources and depth of resistance from all change populations.
  - Determine the impact of the organization's history on the change.
  - Assess the impact of the current culture on the change.
- Design and implement a comprehensive change implementation plan.
  - Surface business risks and recommend action plans to reduce the risk.
  - Balance effective use of resources with the requirements of the change.
  - Build a communication plan.
  - Build a learning plan.
  - Build a reward/reinforcement plan.

On the other hand, when we refer to *competencies* for the change agent, that means:

- Demonstrate people and process expertise.
- Develop effective teams.
- Communicate with sponsors, peers, and target populations.
- Transfer problem-solving skills.
- Manage projects.

The role of the change agent is so important that an evaluation may be appropriate. An example of a simple evaluation for a typical change agent may be the following, which consists of several statements. Obviously, these statements are not exhaustive ones but rather common and simple ones. They may be presented in a Likert-scale format (1–6, where 1 = *strongly agree* and 6 = *strongly disagree*).

- Understands the change to be implemented.
- Has successfully implemented change in the past.
- Is trusted and respected by sponsors of the change.
- Is trusted and respected by targets of the change.
- Shows knowledge and sensitivity to both people and process issues in the change.
- Communicates effectively, both in packaging messages and in active listening.
- Builds team and strives for involvement and consensus.
- Exhibits patience in working with sponsors and targets.
- Experienced in the use of the tools of project management.
- Demonstrates a thorough understanding of the change and the change process.
- Builds effective strategies and tactics to support change.
- Understands the targets' issues and resistance.
- Knows what is required of sponsors and can negotiate for that sponsorship.
- Has demonstrated the ability to educate sponsors in their role.
- Balances effective use of resources with the requirements of the change.

It is imperative to understand that the change agent is also a communicator. Therefore, everyone in the organization must take into account what the change is for and what kind of a communicator is needed. Typically, there are four styles:

1. Intuitor

The characteristics of this style are as follows:

- Focus is on long term and big picture
- Can visualize new solutions and new ideas
- Not limited by traditions or old ways of doing things
- Good conceptualizer
- Can develop a sense of structure out of confusing or uncertain situations

The weaknesses may be that they can

- Be overly abstract and theoretical
- Be impatient or unconcerned about practical and business issues
- Appear judgmental and arrogant

With this type of change agent, typical outcomes may be as follows:

Successful	Unsuccessful
Creative	Unrealistic
Idealistic	Impractical
Charismatic	Condescending
Original	Rigid
Comprehensive	Fragmented

2. Thinker

The characteristics of this style are as follows:

- Analytical and logical in viewing problems
- Focus is on observation, proof, and rational processes
- Highly effective organizer
- Skeptical about new ideas and first impressions
- Consistent and reliable with a steady, disciplined follow-through

The weaknesses may be that they can

- Be overly cautious and reluctant to take action
- Be unwilling to try new ideas
- Be unconcerned about people issues

With this type of change agent, typical outcomes may be as follows:

Successful	Unsuccessful
Objective	Unconcerned
Analytical	"Analysis paralysis"
Deliberate	Indecisive
Controlled	Rigid
Task oriented	Insensitive

3. Feeler

The characteristics of this style are as follows:

- Perceptive
- Patient and empathic listener and observer
- Sensitive to political and "unsaid" concerns
- Effective in bringing different people together and developing teamwork
- Can balance objective issues with "gut feeling"

The weaknesses may be that they

- Focus too much on people issues and avoid necessary action
- Are too impulsive and dramatic to be viewed as reliable
- Are preoccupied with politics and intrigue, not business issues

With this type of change agent, typical outcomes may be as follows:

Successful	Unsuccessful
Empathic	Sentimental
Spontaneous	Impulsive
People oriented	Not business oriented
Feeling	Inactive
Persuasive	Manipulative

4. Sensor

The characteristics of this style are as follows:

- Highly focused on bottom line results
- Very energetic and hard driving
- Prefers to manage multiple tasks at the same time
- Generally well organized and pragmatic
- Sets high standards for self and others

The weaknesses may be that they can

- Be too unconcerned with long-range impact
- Act too quickly before adequate planning is done
- Perceive different opinions as resistance or disloyalty

With this type of change agent, typical outcomes may be as follows:

Successful	Unsuccessful
Results oriented	"Ready/fire/aim"
Confident	Arrogant
Task focused	Tunnel vision
Impulsive	Unreliable
High standards	Unrealistic

In this phase we will also determine the degree of risk and the cost of the change to the culture, history, and resistance to the organization, employees, customers, governmental agencies, and society.

After the risk analysis is completed we are ready to plan the actual change; that is, to design the change system, which includes the communication system, learning system, and reward system.

Planning the change obviously is a major component of the entire process, but it is not the entire process. Just because you have planned something does not mean that is executed. Therefore, the next phase is the execution of the plan. That means building the change strategies and tactics of your plan into the organization.

The final phase is the monitoring stage, where observations, measurements, as well as adaptation behaviors are evaluated and adopted as the change is changing the organization.

CHANGE AND SCENARIO THINKING

Perhaps one of the most important issues in any pending change is the prediction of that change. One of the successful strategies in forecasting has been and will continue to be *scenario thinking*. Scenario thinking encourages executives to step into the unknown and imagine a range of possible futures. Scenario thinking is not to be confused with the *Delphi method*, which is a tool for increasing consensus within a large group of people about future expectations. Also, whereas the Delphi method has preset criteria for ending the consensus, scenario thinking is completely free to analyze any issue.

When you ask most executives about risk management, they speak earnestly about things like hedging against currency fluctuation, buying insurance, or securing options to guarantee future price stability. They respond to what they think is a quantitative question with a bevy of quantitative solutions. They assume that you are asking about traditional risk management—the kind most often associated with finance, insurance, law, safety and security, the kind they can model and quantify.

But what if you are really asking about a different type of risk management? And what if you are not asking about the narrow sorts of risk they expect—the kind that can be managed, controlled, and avoided—but the broader types of risk that they have not yet put on their radar, let alone learned how to think about strategically? Quality is one such area. Most people associate it with defects; however, we increasingly associate quality with customer satisfaction. This transition presents problems because customer satisfaction is dependent on personal, individual expectation and it is quite dynamic in nature. Furthermore, each customer defines quality as he or she sees fit.

In 1999, one of Enron's strategic partners asked the question: What if Enron were to fail? At the time, the company was a picture of success: it was the seventh largest U.S. corporation, with more than $100 billion in stated revenue. Nevertheless, in order to sharpen its strategy and explore what risks might lie ahead, the company's executive team developed four scenarios. One of its scenarios, called "Starting Over," anticipated Enron's failure. Although they were never heeded, the executive team had identified risks that no one was considering.

In the months following September 11, 2001, at a time when U.S. government was being challenged in new ways around the globe, a number of leading companies based in the United States conducted a rigorous exercise to imagine how they were perceived outside the country. These companies were concerned about their brands, as well as the security of their plants and personnel in areas where U.S. sentiment was running low, such as the Middle East and even France. Despite the difficulty associated with quantifying brand decay, they built scenarios to understand the brand risks they faced abroad and the opportunities at home. Had they taken a narrow approach to risk and risk management, they would not have considered brand risk, because it is so hard to measure.

Today and in the future, many companies are pondering their strategies for China and other emerging economies. For example, should India, Cambodia, Thailand, China and others be viewed as a center for low-cost manufacturing? Are they fast-growing consumer markets? Are they emerging threats to U.S. national security? Are they growing to the point where one of those countries could be a world leader that could raise the global trade? Right now, in executive suites across the country, scenarios are being used to understand the many potential outcomes for the uncertainty of such questions that China faces, without presupposing them as risks or opportunities. Without this approach, what might these companies have missed about what is coming?

Though there will always be a need for traditional risk management, none of the strategic learning highlighted by these examples—anticipating the future demise of an industry giant; understanding the impact of September 11 on IBM or Coca-Cola's brands; exploring the range of effects that China might have on Wal-Mart, Philips, or Unocal in 10 to 15 years—would be achievable through the traditional tools and perspectives of risk management.

In the past few years, we have come to recognize that change is coming fast in everything that we do. We also are recognizing that the definition of risk management for organizations has broadened. In fact, that definition has been expanding beyond the tangible and quantifiable issues for which executives possess well-honed tools to the less tangible and more qualitative forms of risk that few have learned how to anticipate, recognize, or respond to proactively. In this new world of change and of risk, the bounded definition to which most executives subscribe is of limited use— not only because it causes them to miss potentially major disruptions but also because it blinds them to the considerable opportunities that come when risk is well (or even eagerly) anticipated.

Indeed, the classic separation in organizations between risk managers and risk takers—the investors, entrepreneurs, innovators, and strategists—overlooks the fact that these latter opportunity-minded professionals manage risk, too, just in a different way. Their job is to look for ways to create value by exploiting uncertainties in the business environment—in other words, to take reasoned risks based on a point of view about where the world is going, despite the uncertainty they face.

An example is the future of the automotive industry and the environmental industry. Companies are proposing high-mileage vehicles, hybrids, electric cars, fuel cells, and other alternatives without really knowing the ramifications and consumer acceptance of those alternatives. As good as these items may sound, they are indeed foolish and not doable in the foreseeable future, primarily due to high cost. If automotive companies cannot sell their vehicles at an average cost of about US$25,000, how will they be able to sell vehicles at an average cost of US$45,000? Why would anybody buy a vehicle that will have a range of about 40–50 miles, a cost of about US$40,000, a replaceable battery that costs about US$10,000, and a life expectancy of about 12–15 years?

In the case of the environmental industry, more hoaxes are being presented to the public without honest discussions and appropriate scientific research. Let us start with global warming, which, of course, is the driving force for many environmental initiatives. The carbon footprint is the culprit. Allegedly, man is the worst enemy. We forget that a typical volcano, which occurs naturally, contributes more carbon than humans do. Wind is another issue. Indeed, wind provides power, but it is not consistent. The ancient Greeks and Middle Age Holland used wind power generation but gave it up because they found that it was not effective for substantial and across-the-board applications. Hydropower, sea waves, natural gas, and many other alternatives exist, but hardly anyone speaks openly about them. It seems that politicians and extreme groups are focusing on environmental issues only from the perspective of reducing fossil fuels, disregarding other opportunities with much more return on the original investment.

Executives, in order to better position themselves to see the risks and opportunities ahead of them, need to integrate their risk management and risk-taking sides—within the context of social responsibility, within themselves, and within their organizations. This will require them to adopt a new attitude—to take a more strategic view of risk that incorporates exploring and managing uncertainty.

This is precisely what scenario thinking is all about. It is an approach for managing risk in a changing environment. The new mind-set for risk management requires a set of approaches that fully embrace complexity and ambiguity and allow for a more balanced view of both the risks and opportunities in the business environment. Scenario thinking is a key part of this toolkit because of the unique ways in which it allows leaders to explore and exploit the unknown and because of its ability to enable action in the face of uncertainty. A quick look at some of the main components of scenario thinking reveals how this can be achieved.

ARTICULATING THE OFFICIAL FUTURE

Scenario thinking exercises generally begin with a rigorous fact-finding phase, in which the official future is articulated. Rather than conducting research on how a company is performing, or what the future is likely to be, the research builds a story describing the assumptions that management has about the future based on the management's observable actions.

Official future exercises tend to be revealing, because they make explicit assumptions that otherwise tend to go unnamed. In creating an official future for the "Big Three," one might suggest that even though the public at large may still want individual cars, the challenge is to be ecofriendly, cost-effective in both original price and life cycle costs, as well as high performing. The reader will notice that some of these objectives are contradictory, and this is where the opportunity exists for breakthroughs. With scenario thinking these alternatives may be analyzed, evaluated, and decided upon. Similarly, an official future exercise for an oil company would suggest that a growing business opportunity is emerging from the transition to a diversified fuel economy, integrating hydrogen, solar, and wind energy with traditional fossil fuels. By analyzing today's actions and commitments, thoughtful analysts infer what management must believe about the future.

This official future analysis is a foundation for exploring what is known and what is unknown about the future within an organization: it provides a snapshot of management's perceptions. Most of the time, executives disagree with the official future when it is presented to them—that is, they reject the underlying assumptions that they themselves are making implicitly as they execute their companies' strategies. When presented systematically with a clear view of their assumptions, they see for themselves where the logic breaks down. What is most startling about this is that they then realize, without quantitative data and without sophisticated risk models, that they are not fully accounting for the risks and opportunities their companies face.

Scenarios—plausible future narratives—allow for a thorough exploration of future risks and opportunities. It is important to note that scenario thinkers create multiple futures, rather than just one. This allows for a more complete exploration of the future that is not tied to a specific set of assumptions about how uncertainties will unfold. By demanding something different from a prediction, complemented by a best-case and worst-case scenario, scenario thinking gets leaders to consider alternative worldviews.

Scenario thinkers begin at the same place as traditional risk managers, by skillfully making an inventory of what is known about the future. After exploring issues such as demographics as well as aspects of industry structure and customer behavior, scenario thinkers turn to the unknown, the unknowable, and the perceptions that should be challenged based on the official future exercise. Following a rigorous analytical process aimed at articulating the range of uncertainties an organization could face and all of the relevant outcomes, scenario thinkers design a number of future narratives.

The process of developing multiple scenarios helps to increase the possibility that executives will not be surprised, because it allows them to rehearse multiple unique futures (contingencies). The process also grounds decision makers in the reality that, in most circumstances, they cannot accurately predict the future. Rather than falsely assuming that one outcome will happen, leaders then understand that they must make decisions in light of the true uncertainty they face.

In order to explore the range of futures the organization may face, most executives need to broaden their context beyond internal uncertainty. They must always be aware of what is happening in their industry at large as well as be able to understand the broader contextual environment.

The broader context of social, political, economic, environmental, and technological domains offers tremendous opportunities and threats to most organizations. Often, though, risk management tools do not provide a sufficient connection between immediate decisions impacting an organization and the external environment. Because this broader context is often less known and less controllable, it is more foreign to executives and, therefore, is ignored. Most scenario thinking processes devote substantial resources to exploring the many ways in which the broader context could evolve and the tactical decisions it imposes on an organization.

As we have mentioned, scenarios are written as plausible stories—not probable ones. Traditional risk management is based on probabilities, actuary tables, and other known and measurable quantities. But scenarios are intended to provoke the imagination and provide a more comprehensive view of risk, so that the results of the pending change can be embedded in critical strategic decisions. If scenario thinkers focused only on what was likely or predictable within some reasonable confidence interval, they would not identify as broad a range of risks or as many opportunities as they do. Because scenarios aim to highlight possibilities that could have a meaningful impact on the organization, plausibility is a more relevant criterion than a specific prediction.

In the past, scenarios were often used primarily to uncover new risks and identify new opportunities; scenario thinking frameworks are currently being integrated into organizational decision-making processes. Companies conduct thorough analysis to isolate scenario-specific actions from robust ones—actions that are relevant in any foreseeable future. By carefully ranking the importance of strategic choices that an organization faces in various scenarios, analysts can identify which are most likely to be important in all situations and which are only relevant given specific future conditions. Sophisticated models are then built to track the progression of contextual conditions and a company's corresponding strategy.

Obviously, one of the challenges in implementing scenario thinking effectively is to maintain the new risk management mind-set throughout the process. It is sometimes tempting to demand that scenario thinking processes offer the type of outputs that traditional risk management approaches offer, such as predictions, hedging strategies, and insurance plans. But scenario thinking has a more ambitious goal: it enables a more complete view of the risks and opportunities a company faces and offers a framework for engaging those findings into the strategic planning process.

The new mind-set for risk managers requires rituals and approaches that are deeply embedded in the scenario thinking process: a capacity for learning, an appreciation of uncertainty and ambiguity, an understanding of the value of strategic conversation, and a willingness to explore uncharted territory. Increasingly, executives are appreciating that the changing nature of risk requires approaches that may initially be uncomfortable but over time turn out to be more effective in embracing the unknown.

The boundaries of risk management as they relate to change are continually being challenged. With faster and more advanced computing power, we can expect improved models for understanding and valuing risks associated with highly complicated processes, such as climate change. As the definition of risk management broadens, we can also expect to see the wider use of approaches such as scenario thinking that encompass the many facets of risk.

We can also expect to see wider and more integrated use of scenario thinking within organizations, because managing risk and uncertainty is becoming everyone's business. After all, there is no guarantee of success for any future action for any pending change. The best we can do is to plan for the risk of failure or to minimize failure by defining and planning for mediating circumstances. Globalization is more or less forcing us to think in terms of new economic and management models: product research and development now have distributed workforces; market research is more complicated and far-reaching; quality is constantly redefined and international standards are being adopted in just about every industry; and manufacturing faces new cost models. It is hard to think of a part of the company that is not potentially implicated.

Skilled scenario thinkers take a holistic view of risk and so, too, should organizations. Scenarios are indeed a methodology for simplifying the risk and opportunity that a business faces when change is either planned or comes by surprise from outside forces. And scenario planners are quick to warn that not using such a rigorous and progressive risk management approach may be the biggest risk of all.

SUMMARY

In this chapter we have discussed engagement. Specifically, we focus on diffusion of the leader's ideas in the organization using communication skills, understanding change, and scenario planning. In the next chapter we give an overview of teams and their role in the pursuit of improvement.

REFERENCES

Ackoff, R. L., H. J. Addison, and S. Bibb. (2007). *Management f-LAWS: How organizations really work*. Devon, UK: Triarchy Press Ltd.

Ensman, R. G., Jr. (2005). "The top 10 delegation mistakes ... and how to avoid them." *Office Solutions* September–October, 30–31.

f-LAWS. http://www.f-laws.com/content/what_are_f-laws.php.

Karlsson, P. J. (September 25, 2009) http://www.playenglish.net/simple/?t8407.html.

Novations Group (2008). http://www.globalnovations.com/Home.html.

Rizwan, A. (October 22, 2007). http://www.rizwan.in/node/10930.

Sull, D. N. (2005). *Made in China: What Western managers can learn from trailblazing Chinese entrepreneurs*. Boston: Harvard Business School Press.

SELECTED BIBLIOGRAPHY

Argyris, C. (1955). "Top management dilemma: Company needs vs. individual development." *Personnel* September,123–134.

Backlund, P. and D. Ivy. (2004). *Gender speak: Personal effectiveness in gender communication,* 3rd ed. Boston: McGraw Hill.

Bennis, W. (1990). "The manager and the leader." *Training Magazine* May, 25–31.

Bovée, C. L. and J. V. Thill. (2005). *Excellence in business communication*, 6th ed. Upper Saddle River, NJ: Pearson-Prentice Hall.

Burgoon, J. K., D. B. Buller, and W. G. Woodall. (1989). *Nonverbal communication: The unspoken dialogue*. New York: Harper and Row.

DeKlerk, V. (1991). "Expletives: Men only?" *Communication Monographs* 589: 156–169.

Forgas, J. P. (1991). Affect and person perception. In *Emotion and social judgments*, edited by J. P. Forgas. New York: Pergamon Press.

Hocker, J. L. and W. W. Wilmot. (2001). *Interpersonal conflict*, 6th ed. Boston: McGraw Hill.

Jones, M. (2002). *Social psychology of prejudice*. New York: Prentice Hall.

Kelman, H. C. (1958). "Compliance, identification, and internalization: Three processes of attitude change." *Journal of Conflict Resolution* 2(1): 51–60.

Kelman, H. C. and V. D. Hamilton. (1989). *Crimes of obedience: Toward a social psychology of authority and responsibility*. New Haven, CT: Yale University Press.

Kilmann, R. and K. W. Thomas. (1975). "Interpersonal conflict—Handling behavior as reflections of Jungian personality dimensions." *Psychological Reports* 37: 971–980.

Likert, R. (1961). *New patterns of management*. New York: McGraw Hill.

Mager, N. R. and M. A. Rahim. (1995). "Confirmatory factor analysis of the styles of handling interpersonal conflict: First-order factor model and its invariance across groups." *Journal of Applied Psychology* 80(1): 122–132.

Maslow, A. H. (1954). *Motivation and personality*. New York: Harper & Row.

McGregor, D. (1960). *The human side of enterprise*. New York: McGraw-Hill.

Meadows, E. (1999). *The Viking way to infinity*, 4th ed.

Pincus, M. (2003). *Everyday business etiquette*. Bloomington, IN: Authorhouse.

Randall, D. and C. Ertel. (2005). "Moving beyond the official future." *Financial Times* 16 September.

Richmond, V. P. and J. C. McCroskey. (1995). *Communication: Apprehension, avoidance, and effectiveness*, 4th ed. Scottsdale, AZ: Gorsuch Scarisbrick.

Tannenbaum, R and W. Schmidt II. (1958). How to choose a leadership pattern. *Harvard Business Review* March–April, 95–101.

Townsend, P. (1999). "Fitting teamwork into the grand scheme of things." *Journal for Quality and Participation* 25(1): 16–18.

Verderber, K. S. and R. F. Verderber. (2004). *Interpersonal communication concepts, skills, and contexts*. New York: Oxford University Press.

8

Team Building

In the last chapter we discussed how a leader in any organization must engage all stakeholders to achieve a positive result. In this chapter we give an overview of team building and its role in the improvement process.

In no uncertain terms, any quality effort in any organization is a team effort. That is, *Together Everyone Accomplishes More*. It is important to note here that the word itself implies as much, for there is no *I* in *team*.

When one begins to talk about the team experience, right away it must be realized that teams are not inherently there. Rather, they develop from a group environment. Therefore, it is important to understand the group dynamics. Groups form a basic unit of work activity throughout engineering, purchasing, sales, marketing, manufacturing, and so on. Yet, the underlying process is poorly managed.

Mismanagement has existed for many thousands of years. After all, even in the biblical story of creation, God made an individual, and then He made a pair. The pair formed a group, together they begat others, and thus the group grew. Unfortunately, working in a group led to friction. The friction was so strong that it led, first, to the original pair being expelled from Paradise and, second, to the killing of Abel by Cain. There has been trouble with groups ever since.

When people work in groups, there are two quite separate issues involved. The first is the task and the problems involved in getting the job done. Frequently this is the only issue that the group considers. The second is the process of the groupwork itself: the mechanisms by which the group acts as a unit and not as a loose rabble. The group's structure is that participants are equal; however, the leadership is usually governed by the chairperson, a somewhat autocratic approach. The decision is generally made by a majority vote. Without due attention to this process, the value of the group can be diminished or even destroyed; yet with a little explicit management of the process, it can enhance the worth of the group to be many times the sum of the worth of its individuals.

This process is called *synergy*. It is this synergy that makes groupwork attractive in corporate organizations despite the possible problems (and time spent) in group formation, even though it is less than optimum. On the other hand, it must be realized that in a group environment synergy is only as good as the time allowed for the task and at the discretion of the chairperson. After all, the synergy in a group is based on majority vote rather than a true consensus. Typical characteristics of a group are the following:

- Little communication
- No support
- Lack of vision
- Exclusive cliques
- The whole is less than the sum of its parts
- Seeks to hide its identity
- Leaves new members to find their own way but insists on conformity
- Leader manipulates team to own ends

To be sure, a group is better than the dictatorial decision-making process of old. However, there is a better way. That is, the *team approach*. Over the last few years, the team approach has been proven to be very effective. However, it requires constant evaluation and attention, because it is perhaps one of the most valuable resources the organization has. Furthermore, its success depends upon management's encouragement to use teams and the cultivation of the culture to accept the teams and their decisions.

So, what is the transition from a group to a team? First let us talk about what a team is and then about the transition. A team allows for two or more individuals to be working together for a common goal, in one room, on a common project, having ownership of that project and deciding by consensus. Consensus is the process of agreeing on the outcome of the team even though there are some disagreements. In other words, participants can live with the disagreements and call the decision their own. It is this process (consensus) that allows for true open communication, encouragement of participation, mutual support for everyone, and ultimately better performance through innovative solutions. The typical characteristics of a team provide for

- Plenty of opportunity for discussion.

- Plenty of support (from management and participants for each other).
- A process of discovery supported by openness and honesty.
- Tactical and workgroups combining to form a single team.
- True synergy—the whole is greater than the sum of its parts.
- The opportunity for participants to seek and discover their identities.
- New members to be welcomed by showing them existing norms and openness to change.
- The leader to seek team decisions by serving the team as a facilitator for two-way communication.

Now that we have a general idea of what a team is, let us examine the process of getting there. Team building is creating the opportunity for people to come together and share their concerns, ideas, and experiences and learn to work together more effectively to solve their mutual problems and achieve common goals. A team-building program involves putting data gathering, diagnosis, and action planning activities into a framework with action taking and evaluation as follow-up activities. The focus of these activities centers on the following questions:

- What keeps us from being as effective as we could?
- What problems do you experience that we should work on?
- What changes do you feel need to be made to be more effective?

The complexity of team building is always there and demands management's interventions as well as facilitation of the team development process. This process has been defined by Tuckman and Jensen (1977) and may be summarized as follows:

- *Forming:* The initial stage of small-group development is characterized by a movement toward awareness. In the process of forming, members attempt to become oriented to the tasks, as well as to one another. (This is the first attempt to become a team.) The amount of information available and the manner in which it is presented is critical to team development. Resolving dependency issues and testing are the major relationship behaviors. Understanding leadership roles and getting acquainted with other team members facilitate team development in this stage.

- *Storming:* When orientation and dependency issues are resolved, conflict begins to emerge, signaling the second stage of team development. The storming process involves resistance or emotional responses to task demands and intrapersonal hostility in relationships. Team members engage in behaviors that challenge the group's leadership or they isolate themselves from team interaction. If conflict is permitted to exceed controllable limits, anxiety and tension permeate the team. If conflict is suppressed and not permitted to occur, resentment and bitterness result. This can encourage apathy or abandonment of the team. Although conflict resolution often is the goal during the storming stage, conflict management is generally what is achieved. In fact, conflict management is a more appropriate goal because it is desirable to maintain conflict at a manageable level to encourage the continued growth and development of the team.

- *Norming:* The third stage of small-group development into a team is norming and is characterized by cooperation. The dominant task themes are communication and expression of opinions. Sharing of information and influence promotes cooperation and synergistic outcomes. Cohesion is the relationship theme. A blend of harmony and openness is created by the work effort, which increases morale and team-building efforts. Group unity develops, and shared responsibilities increase, typically leading to decision making by consensus and democratic leadership styles.

- *Performing:* The fourth stage of small-group development is evidenced by productivity. Performing encourages functional role relatedness. The task theme is problem solving. Team effort is made to achieve team goals. Team members provide valuable contributions by assuming appropriate roles that enhance problem solving. The relationship theme is interdependence; it is the basis for any successful team effort and requires team members to be simultaneously highly independent and highly dependent.

Bringing the team to a conclusion creates some apprehension; in effect, it is a minor crisis. Therefore, as the team approaches its conclusion the tasks undertaken will be less in both quantity and complexity. If the team has been responsible for its own functioning but now seems to be unable or unwilling to continue to do so, the appropriate leadership behavior would be to change to the participating style. Because the adjourning stage of small-group development centers on separation, grieving and leaving behaviors are typical. The termination of the group is a regressive movement from giving up control to giving up inclusion in the group (Schutz 1958). On the other hand, because the group at this stage has been developed into a full-fledged team, the transition is more fluid as well as manageable.

This is so because the participating style facilitates the task termination and disengagement process. A low task orientation allows team members to become actively involved in the team's conclusion and the leader provides high relationship support to combat reluctance to leave and the desire to remain within the safe, predictable structure of the team. If the crisis of termination and disagreement were to persist, the leader would match the decreasing task maturity level by regressing to the selling style and increasing task-directive behaviors.

In most organizations, the main reason for placing a strong team is the recognition of the need for interdependence (Weiss & Molinaro 2005). Most jobs require bringing together the talents of many departments or specialists, or other resources, in a collaborative effort. If the manager is clear about this kind of need, then the next step is to recognize those behaviors or issues that can block and distort the effectiveness of the system (Lencioni 2002). Collaborative behavior is often obstructed by the following:

- Varied perceptions of the task to be performed and the goals.
- Lack of clarity about roles, responsibilities, and authorities.
- Lack of effective means of planning, problem solving, and decision making.
- Inappropriate win–lose relationships.
- Dysfunctional agreement among team members: Each member operates on their own as a result of a poor leader who lacks skills, knowledge and confidence (Lencioni 2005).
- Different feelings of equality of membership or influence in the team; feeling in or out of the group.
- Dependent or rebellious attitudes toward authority.
- Difficulties in interpersonal relationships.
- Financial or other rewards that seem unfair.
- Inability to manage the inevitable conflicts between groups.

Though teams are very useful in many applications, because they are optimistic in nature and forward looking, there are certain situations in which teams should not be used. Some of these situations and conditions are as follows:

- Teams should not be used where the diagnosis of the system indicates that team building is not a major concern but the need exists for job enrichment, new equipment and machinery, technical or skills training, etc.
- Teams should not be used if it is not necessary to the results and the effectiveness of the group that they work together on common goals and objectives.
- Teams should not be used if the other parts of the system are likely to undo, or prevent, the changes the group determines to be desirable.
- Teams should not be used if there is no chance for dialogue or negotiation with the rest of the organization, then team building can generate aspirations and enthusiasm that can only lead to increased disappointment.
- Teams should not be used unless the group really has the opportunity to influence its own future.
- If a decision has been made to phase out a group, it is not likely to be helped by team building.

THE LIFE CYCLE OF THE TEAM

Lee describes a team is a living organism (1986, 1996). I have heard it said that one of the great breakthroughs of the 1950s was that management consultants became aware of management teams as entities. Since then, managers and organizational development professionals have devoted enormous efforts to develop healthy, effective teams and to help team members work smoothly together.

My own association with team dynamics has been intensely practical. I have been involved with several social movements, several project teams, and many business organizations. In the process, I have participated in the birth, growth, maturity, decay, and death of many teams. Birth, growth, maturity, decay, and death serve vital purposes in our individual lives and for the entire human species. No condition is superior to the others, and only death is permanent.

Without decay and death, our world would overcrowd yet more quickly and our social systems would ossify. Those in power would remain in power decade after decade while the rest of us followed orders. Everyone would eventually be bored to death with life. Just look at China's government, where the people who governed it in the 1940s were still in control 50 years later. China's stagnant political life could not begin to renew itself until its old guard had passed away.

Death is nature's way of making room for the new and innovative and keeping life interesting! The prospect of death instills in each of us an entrepreneurial sense of urgency about life (Lee 1986, 1996).

Likewise, the birth, growth, decay, and death of an executive team serve critically important functions for the business as a whole and for team members. I will describe the values and drawbacks of each phase of the life cycle both to corporate vigor and to individual growth. I will show how attempts to maintain a highly effective, highly cohesive management team undermine both the health of the company in which operates and the personal growth of the individuals who are part of that team. It would be better for all concerned to hasten the death process rather than fight it.

The Life Cycle of Executive Teams

The period from birth to maturity is typically 2 to 3 years. Maturity to decay may take 2 to 5 years. Decay to death takes less than a year and is triggered (usually) by a catastrophe the team produces.

An executive team is formed to achieve specific strategic business objectives within a few years. (It cannot be overstated here that some teams are of much shorter duration, depending on the scope of the project.) When the selection process is done correctly, team members are chosen based on their individual abilities to contribute to achieving those objectives. In the first few months of the team's life, its cohesiveness and effectiveness are low. There is much uncertainty about how the team will work together, what each member will contribute, and how the team will fare as it interacts with the outside world. Team members are barely committed to the team and are still strongly immersed in their external environments. There are abundant challenges and healthy doses of the unexpected and fun. There is uncertainty and anxiety about whether or not the team will succeed. The team's energy is concentrated on future successes. Each team member contributes his own stimuli and information from his reality outside the team (Lee 1986, 1996). This is the team's childhood, a time of maximum learning by team members and maximum sensitivity to the world outside the team.

As team members learn from one another and take successful actions together, the team's effectiveness and cohesiveness increase. This increases members' enthusiasm and commitment to the team. For a while there is a positive feedback loop in which success increases cohesiveness, which increases effectiveness, which generates more success. This is the team's adolescence.

Eventually the team accomplishes its first major success, the strategic objective for which it was formed. That strategic success marks the point at which the team is considered to be highly cohesive and highly effective. But cohesiveness has a dark side: lack of openness to the world outside the team or to new team members. Success also has its dark side. The team changes its attitude about its relationship to the outside world. It succeeded; therefore, it has the formula. It loses the anxiety and sensitivity to the external environment that contributed to its success.

The team also develops a team memory based on past successes and previous communications. The team memory now defines each member's role, the team's knowledge of the outside world, and how the team operates in that world. The team memory enables the team to perform like an experienced adult, able to quickly handle challenges in previously learned ways. But the team succeeds only as long as the team memory of how things *were* accurately reflects how things *are*. When the outside world changes— for example, in customer requirements, competitors' innovations, or new technologies—the members of a highly cohesive and highly effective team usually do not respond. They continue to see the world through the team memory and act accordingly. After all, that behavior was successful!

Once the team becomes highly effective and highly cohesive, communication of new information between the outside world and the team and among team members deteriorates. The team memory freezes and becomes increasingly detached from the present reality. Team members no longer listen to one another because they already know what to expect. They become bored with their predictable roles. Sooner or later, the team makes decisions that fail to meet members' needs or fail in a changed external environment. The decay process is under way.

After decay becomes well established, some CEOs seek outside help to restore their teams to peak performance. The team members are highly sensitive to their own isolation within the team and remember a team past in which things were much better. Consequently, the restoration efforts tend to focus on communication and cohesiveness. Sometimes these efforts temporarily slow the decay process. Usually they have little impact, especially when the CEO exempts himself or herself from them.

Loss of effectiveness (typified by one or more failed decisions or projects) eventually overcomes the exaggerated management energy committed to cohesiveness, and the team disintegrates. Disintegration (death) frees team members to participate in new teams where they can renew their enthusiasm, develop new personal relationships, and revitalize their atrophied learning processes. Disintegration of the old team also makes room for a new leadership team, one that is able to start out anchored in the real world, ready to deal with things as they are, not as they used to be.

A project team and an executive team start life in much the same way. The significant difference is that a project team is disbanded when it achieves its initial strategic success. Project team members are rewarded, but one of the rewards is not continued employment. Executive team members expect continued employment in return for past success. (Some material in this section have been adopted and modified from Lee 1986, 1996).

CROSS-FUNCTIONAL TEAMS

A cross-functional team is a team that is made up of individuals from different functions, such as production, purchasing, quality, engineering, and so on. However, as strong as their input is for a particular project, the team may suffer in their output. Some of the factors that lead to poor performance are poor selection of team members and miscommunication among cross-functional team members.

For the selection of the team members, a careful analysis of the project must be understood and the appropriate qualified individuals must be selected. Some of the reasons for poor communication may include the following:

- Lack of appreciation of the contributions of other functions. For example, in telecommunications projects, some engineers do not value the input provided by human factors psychologists.
- Old-fashioned turf battles. Some departments play out their competitive games on the field provided by the cross-functional team.
- Different jargon. For example, line department users often do not understand the terminology and technology employed by computer programmers.

- Different work orientations. For example, researchers tend to take a long-term view and have an informal work climate; operations people are more short term and formal; salespeople are usually informal and have a short-term focus. Though one may argue with these generalizations, it is clear that each department or function develops it own work style, which may clash with other styles from other functions.
- Different degrees of interest in the team's outcome. Some cross-functional team members are simply more interested in the team's purpose and may have more to gain from a successful outcome. In one government agency, team members from one bureau have more interest in the outcome of the team because it affects their client group more than it does the other bureaus represented on the team.
- Mistaken goals. Some team members mistakenly see harmony as the goal of cross-functional teamwork. As a result, they are afraid to express a contrary point of view for fear that it will destroy the positive feelings among team members. The new result is a false consensus and a less than satisfactory outcome.

Though these factors explain lack of trust and communication on cross-functional teams, they do not excuse it. Members of cross-functional teams are there because they have something to contribute. They must be allowed and even encouraged to share their ideas, information, and opinions without restrictions. Open communication is an absolute requirement for successful cross-functional teamwork. The concept of the cross-functional team is that the outcome—the product, the system, the service—will be better because it has been created by the combined expertise (synergy) of people from a variety of functions. Viewing a problem or an issue from many vantage points is the strength of the cross-functional team. However, the value of divergent views can only be realized when there is a free flow of information.

LEADERSHIP AND PARADIGMS

All teams have a leader. A leader is a person you will follow to a place you would not go by yourself. You manage within a paradigm (paradigm enhancement). Give a manager a good system (the rules, the guiding

TABLE 8.1

Comparison Between a Manager and a Leader

Manager	**Leader**
Administers	Innovates
Has a short-range view	Has a long-range perspective
Asks how and when	Asks "What?" and "Why?"
Has his eye on the bottom line	Has his eye on the horizon
Accepts the status quo	Challenges the status quo

principles, the system, the standards, the protocols) and he or she will optimize it. We spend 90 percent of our lives doing just this—evolving.

In the team environment, the leader leads between paradigms (paradigm shift). This shift is the change from the status quo. It is the leaders, with their intuitive judgment to assess the risk, who determine that shifting paradigms is the correct thing to do and instill courage in others to follow them. This kind of change occurs during less than 10 percent of our lives.

Paradigm shifting without the follow-on skills of paradigm enhancing leaves you vulnerable to the paradigm pioneers who practice total quality. Paradigm enhancement without the skills of paradigm shifting will lead you to continually improve obsolete products and services. Nobody will buy obsolete excellence. It is very fundamental here to recognize the differences between managers and leaders. A summary of those differences has been articulated by Bennis (1990b) and is shown in Table 8.1.

WHAT IS THE ROLE OF A CONSULTANT AND/OR A TRAINER?

In any team environment, sooner or later, a consultant and/or trainer will be involved to some degree. However, the question is always the same; that is, What is the role of the consultant and/or trainer in the team process? Understanding the theory of small-group development and its implications for leadership behavior can be a valuable tool for the consultant or trainer. The behavioral themes of each stage of team development offer insight into what happens naturally in a group; patterns that deviate from these themes suggest problems and a need for intervention. Such patterns might include moving too fast, skipping stages, focusing only on task dimensions, and blockages or fixations in particular stages. For example, a foreman who is eager to accomplish as much as possible to impress superiors may spend little or no time on the forming process (stage one) in order to get directly to work on the task. This undoubtedly will result in confusion and misunderstandings among his or her group (not a team yet) of employees, thus hindering both team development and task accomplishment. A manager who dislikes dealing with conflict may make quick, authoritative interventions or delegate decisions to the group, thus abdicating his or her role in helping to manage the group's conflict. The group is likely to view such behavior as unreasonable as well as unhelpful. Similarly, a manager who is very task oriented may be unaware of relationship issues and fail to deal at all with dependency or hostility early in the development of a staff group. The resulting work atmosphere would be one of independent, individual effort in a rigid, inflexible pattern. Conversely, if a task force leader works too hard at building relationships, the team will develop problems during the norming process (stage three) because of the leader's need to seek consensus on minor details.

Mismatches between the stage of small-group development and leadership style often create serious problems. A leader's reluctance to change styles, generally because he or she is comfortable with only one or two styles or has developed skill in using only one or two styles, limits the leader's effectiveness and the team's chances for success. For example, a supervisor who favors the participating style may have difficulty initiating team work. Although more experienced workers may assume responsibility, the more inexperienced ones will flounder or withdraw from team action unless more task direction is provided. While seeking the security of task direction, members of new teams are likely to perceive leader behavior that is highly relationship oriented as inappropriate; in fact, they may be suspicious of it. In another example, a company president who has used the speciality-task, authoritive style in establishing the firm may be very helpful to new project groups in their early stages. But if the president cannot change his or her leadership style once the teams become more functional, this domineering and crisis-oriented manner will prevent open discussion and true efforts at consensus.

Low morale, apathy or resentment, and low productivity may cause groups to finish prematurely. In contrast, but equally ineffective, is the manager who has been encouraged to delegate more as the workforce becomes more cohesive and productive but who delegates only those tasks that he or she personally dislikes doing. As the team members begin to perceive this, and especially if the manager also forgets to check on progress and provide praise or other rewards for superior work, it is unlikely that the tasks will be completed well or on time.

The consultant or trainer may want to examine the motivation and power issues of the leader and the group to ascertain the impact of these on performance behavior. These concerns also provide excellent topics for staff training and development programs in organizations and for trainers and consultants who work with teams and their leaders and who want to increase the timeliness and effectiveness of their interventions.

THOUGHT-PROVOKING OBSERVATIONS

1. When a team is formed, it focuses on the future. Once it succeeds, it focuses on the past. Team members are usually selected based on how they will contribute to the team's strategic objectives. Once the team attains its first strategic success, however, a member of an executive team gets to stay on the team as a reward for the team's success. That member may not be appropriate for the future challenge. (An executive team has to fail repeatedly and miserably before team members are disenfranchised.) IBM lost most of the PC market (new challenge) because its key business decisions were made by people who succeeded with mainframes (past successes).

2. Success breeds failure. In business and in sports it is difficult for a team to repeat its success. A study of management teams found that most successes are followed by major failures. For example, the IBM PC (success) was followed by the PCjr (failure). Apple II (success) was followed by Lisa (failure)! Apple MacIntosh begat Newton! There are almost no "three-peats" in sports or business.

3. Failure can breed success. Norman Schwartzkopf and Colin Powell endured the failure of Vietnam. They learned from that and fought Desert Storm with the wisdom and anxiety that Vietnam fostered. I would not select Norman Schwartzkopf to lead another battle because he succeeded in the last one. He might have too little sensitivity to changed circumstances and, thus, may be inclined to repeat past actions.

4. Term limits of no more than 8 years for executives and executive teams would improve business effectiveness more than any other management change. History has demonstrated time and again that

a true leader can only lead change in the first 2 years of tenure. After that he or she can only maintain a past direction, regardless of any change in his personal vision!

If the management goal is predictable, consistent responses to a changing world leave a team in place indefinitely (the Pope and his cardinals, China's leadership, judges). If the goal is innovative change and consistent successes in a dynamic environment, then CEOs and their executive teams should serve no longer than 8 years! We have been wise enough to put an 8-year term limit on the president of the United States (and, thus, his cabinet). We have not done so for Congress or business executives yet. An opportunity awaits management gurus and boards of directors. Of course, I am not holding my breath.

CONCLUSION

A group becomes a team primarily because there is a task that requires individuals to interact with and influence each other in order to accomplish that task. However, a highly effective, highly cohesive team is a transitory state in a dynamic process. Therefore, as Bennis (1990a) appropriately declared, the team development process must be

- A way of life.
- The responsibility of every team member.
- A continuous process.
- About developing a clear and unique identity.
- Focused on a clear and consistent set of goals.
- Concerned with the needs and ambitions of each team member recognizing the unique contribution that each individual can make.
- Aware of the potential of the team as a unit.
- Results oriented.
- Enjoyable.

and not

- A short term, flavor of the month.
- Imposed without regard to peoples' feelings.

- Spasmodic.
- Reserved for only some members of the team.
- An excuse for not meeting personal responsibilities.
- A process in which actions clearly contradict intentions.
- Seen as a chore.

Business management will improve significantly when executives respect the values of that process and work with its dynamics. To actually go through the formation, some basic characteristics and assumptions are necessary. They are as follows:

- The team exists and is always part of a large system.
- The environment (demand system) must be understood in order to define the team's goal(s) and task(s) related to that goal.
- The demand system is not fixed—it is always dynamic.
- A contingency set of procedures must be adopted in order to define and accomplish tasks that result from the complex and changing nature of the demand system.

It must be emphasized that there is no one best way to organize a team. All teams do spend time and energy dealing with the factors discussed earlier. Stamatis (2002) elaborated on and emphasized some of the key points that are essential for all teams and provided additional references for further study.

Some advocates recommend a democratic leadership style in order to allow and encourage the individual interaction necessary for team development and eventual shared leadership. The application of situational leadership to small-team development suggests that this style works best during the latter stages of team life (norming, performing, and adjourning) and is not effective during the early stages forming and storming. Fiedler (1967) suggested that a democratic style of leadership is most appropriate for moderately structured teams and that highly unstructured or structured teams profit most from a directive style of leadership. Thus, as the team is forming, uncommitted and uncertain of itself and its task, the more directive authoritative style (a style that provides specific instructions and closely supervises performance) provides instruction, direction, and structure under the guidance of the leader.

During the storming stage, team members question things, ask for clarification, and begin to develop trust. The supportive selling style (a style that explains decisions and provides opportunity for clarification) encourages group member involvement and reinforces performance while still providing some direction and impetus to the group's activities. As group members gain experience in working together and in working on the task at hand, they are able to handle more responsibility, and a more democratic style of leadership can be used (Hersey and Blanchard 1982). In fact, they do become teams. The participating (share ideas and facilitate in making decisions) and delegating (turn over responsibility for decisions and implementations) styles of leadership offer opportunities for team members to begin norming and performing with shared and democratic leadership. Eventually, even supportive relationship behavior from the leader is reduced as it is replaced by individual pride and self-motivation.

Finally; as termination of the team approaches and the crisis of separation ensues, supportive leader relationship behavior again is increased to help the team to deal with termination issues. It should be obvious by now that matching the appropriate leadership style with each specific stage of team development not only answers the needs of the team members and facilitates team action and development but encourages individual members and the team as a whole to increase in task maturity. However, the ability to diagnose the stage of development of the team and the knowledge of which style to use is not enough. Attaining skill in actually using each of the four leadership styles and the ability to change styles as the team becomes more mature or regresses are necessary developmental steps for any leader, manager, trainer, or consultant.

SUMMARY

In this chapter we have focused on the most precious item in all organizations: the team and its development as well as its contribution to the organization via the concept of synergy. In the next chapter we will discuss the quality operating system and how this system brings the quality initiatives up for discussion.

REFERENCES

Bennis, W. (1990a). *On becoming a leader*. New York: Perseus Press.

Bennis, W. (1990b). "The manager and the leader." *Training Magazine* May, 25–31.

Fiedler, F. (1967). *A theory of leadership effectiveness*. New York: McGraw-Hill.

Hersey, P. and K. H. Blanchard. (1982). *Management of organizational behavior: Utilizing human resources*, 4th ed. Englewood Cliffs, NJ: Prentice-Hall.

Kulkarni, V. Building a great team. http://www.scribd.com/doc/14973964/Building-a-Great-Team-116. See also www.hrfolks.com.

Lee, E. (1986, 1996). Life Cycles of Executive Teams: Work with nature, don't put senile teams on life support. 9351 Holt Road, Carmel, CA 9392: http://www.elew.com/life-cycl.htm.

Lencioni, P. (2002). *The Five Dysfunctions of a Team: A Leadership Fable*. San Francisco: Jossey Bass.

Lencioni, P. (2005). *Overcoming the Five Dysfunctions of a Team Workbook*. San Francisco: Jossey Bass.

Schutz, W. D. (1958). *FIRO: A three dimensional theory of interpersonal behavior*. New York: Holt, Rinehart and Winston.

Stamatis, D. H. (2002). *Six Sigma and beyond: Foundations of excellent performance*. Boca Raton, FL: St. Lucie Press.

Tuckman, B. W. and M. A. C. Jensen. (1977). "Stages of small-group development revisited." *Group and Organization Studies* 2(4): 419–427.

Weiss, D.S. & V. Molinaro. (2005). The leadership gap: building leadership capacity for competitive advantage. Mississauga, Ontario: John Wiley & Sons Canada Ltd.

SELECTED BIBLIOGRAPHY

Bales, R. F. (1953). "The equilibrium problem in small group." In *Working papers in the theory of action*, edited by T. Parsons, R. F. Bales, and E. A. Shils. New York: Free Press.

Bales, R. F. (1961). "Task roles and social roles in problem-solving groups." In *Readings in social psychology*, edited by E. Maccoby, T. Newcombe, and E. Hartley. New York: Basic Books.

Bennis, W. G. and H. A. Shepard. (1956). "A theory of group development." *Human Relations* 9: 415–437.

Bion, R. W. (1961). *Experiences in groups*. New York: Basic Books.

Blair, G. M. (1995). *Starting to manage: The essential skills*. Piscataway, NJ: Chartwell-Bratt (UK) and the Institute of Electrical and Electronics Engineers.

Braaten, L. J. (1975). "Developmental phases of encounter groups and related intensive group: A critical review of models and a new proposal." *Interpersonal Development* 5: 112–129.

Hare, A. P. (1976). *Handbook of small group research,* 2nd ed. New York: Free Press.

Kormanski, C. (1982). "Leadership strategies for managing conflict." *Journal for Specialists in Group Work* 7(2): 112–118.

Lacoursiere, R. B. (1974). "A group method to facilitate learning during the stages of psychiatric affiliation." *International Journal of Group Psychotherapy* 24: 342–351.

Mann, R. D. (1967). *Interpersonal styles and group development*. New York: John Wiley.

McGregor, D. (1960). *The human side of enterprise*. New York: McGraw-Hill.

Miles, M. B. (1981). *Learning to work in groups,* 2nd ed. New York: Teachers College Press.

Mills, T. M. (1964). *Group transformation*. Englewood Cliffs, NJ: Prentice-Hall.

Runkel, P. J., M. Lawerence, S. Oldfield, M. Rider, and C. Clark. (1971). "Stages of group development: An empirical test of Tuckman's hypothesis." *Journal of Applied Behavioral Science* 7(2): 180–193.

Schutz, W. D. (1982). *The Schutz measure.* San Diego, CA: University Associates.

Slater, P. E. (1966). *Microcosm: Structural, psychological and religious evolution in groups.* New York: John Wiley.

Spitz, H. and B. Sadock. (1973). "Psychiatric training of graduate nursing students." *New York State Journal of Medicine* June, 1334–1338.

Stanford, G. (1977). *Developing effective classroom groups.* New York: A&W Publishers.

Thelen, H. and W. Dicherman. (1949). "Stereotypes and the growth of groups." *Educational Leadership* 6: 309–316.

Tuckman, B. W. (1965). "Developmental sequencing small groups." *Psychological Bulletin* 63: 384–399.

9

Quality Operating System

So far we have discussed the need for quality and some key items of concern for the quality discipline in the twenty-first century. Though these are indeed important, we must also focus on how we bring all these together. In this chapter we will address the quality operating system (QOS), which does precisely that.

A QOS is a systematic, disciplined approach that uses standardized tools and practices to manage business and achieve ever increasing levels of customer satisfaction. A generic model of a QOS is shown in Figure 9.1.

As complicated as this definition sounds, in fact it is not. All organizations have some type of QOS, but it may be known as:

TQC = Total quality control
TQM = Total quality management
TCS = Total control system
QMS = Quality management system
QRS = Quality reporting system
QIP = Quality improvement process

In essence, QOS is a system that helps organizations to improve both internally, by identifying appropriate metrics for improvement, and externally, by focusing on the customers and their needs. In more generic terms, QOS is

A way of doing business consistently; a systematic approach that focuses on both common and assignable causes to problems; used in all areas of the organization with a focus on improvement; a way to correlate process measurables to customer measurables; a way to implement preventive actions; it is a system that encourages data utilization for decisions and above all it

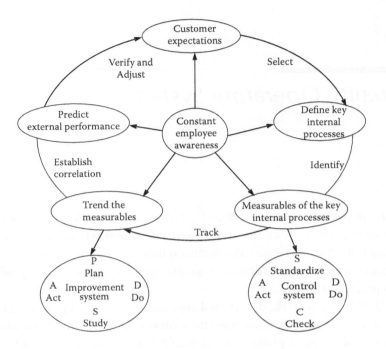

FIGURE 9.1
Generic overview of a QOS model.

forces the organization to work as a team rather than individual entities. QOS is an optimization approach to management rather than a suboptimization for certain departments at the expense of others (Ford 2008, p. 4).

As important as the QOS system is in any organization, it must be recognized that unless management believes in and exercises empowerment at the source, QOS cannot be implemented. *Empowerment*, of course, is personal ownership and commitment to take initiatives aligned with the objectives of the organization.

Let us look at the components of the model itself for a better understanding. We start with customer expectations. It is very easy to conceptualize what the customer wants, but in order to know for sure the organization must know who the customer is. Any organization has many customers and everyone has a unique situation and/or requirement for their product or service. The customer wants a defect-free product, but that usually is a minimum requirement. Some other requirements may be value received, timing (was the product delivered in a timely fashion?), among many others. These, of course, depend on whether or not the customer is perceived as an immediate, intermediate, or ultimate customer. Each one has its own needs and requirements and, depending on the definition of *customer*, the expectations and feedback will differ.

The second component of the model is to select key processes. The operative word here is *key*. This means that only selected processes must be identified and monitored. This, of course, is the Pareto principle (80:20 rule). The process may have many subprocesses, but it is imperative that we identify what the critical processes are that will make the product or service fit the satisfaction of the customer. For example, in manufacturing we may identify assembly, machining, and plating. For staff processes we may identify training, attendance, and accounts receivable.

The third component of the model is to identify the measurables for these key processes. A *measurable* is a way to gage how well a specific aspect of the business is doing and the degree to which the processes of the business meet customer needs and requirements. The focus of this component in the model is to be able to standardize and then control the measurable. We standardize the measurable so that we have consistency and predictability. Once those measurables are standardized, we move into the control mode to make sure that they stay consistent. The two most common types of measurables are the following:

1. Process (production or operation focus)
2. Result (customer focus)

A variety of examples of potential QOS measurables are shown in Table 9.1.

The next key item in the model is tracking the trend of the measurable that was identified in the previous step. This is a very important step in the sense that the control process that began in the previous item, continues; however, the monitoring is more formal and systematic. There are many specific tools that one may use. (See Appendix B and C on the companion CD-ROM, which provide both methodologies and tools.) Here we mention the most common:

- Flowchart—to make sure that we understand the process and that we follow it.
- Check sheet—to make sure that data are appropriately collected and recorded. One must remember that without data, you are only another person with an opinion.

TABLE 9.1

Potential QOS Measurables

Category	Measurable	Quantifier
Generic	Customer rejects and/or returns	Return per 100,000 Parts per million Rejects over time
	Process capability	Percentage of significant characteristics in control Percentage of significant characteristics capable Percentage of control plans completed
	Reliability testing	Mean time between failures Time to failure Field failures
	Absenteeism	Percentage of workforce missing Percentage of excused hours
	Cleanliness	Bathroom Work area Hallway
	Claims	Claims for part 1 Claims for part 2
	Training	Number of hours per person Number of classes Percentage who attended class Productivity improvement
	Cost of quality	Prevention cost Appraisal cost Percentage of sales Internal failure costs External failure costs
Service	Turn days (new business order or repeat order)	Average number of days from point of order entry until shipped
	Quote turn time	Number of quotes completed in 48 hours as a percentage of total quotes processed
	Order acknowledgment	Number of orders that we acknowledge ship dates in less than 48 hours as a percentage of total orders acknowledged
	Order entry	Number of orders entered the same day as received as a percentage of total orders received
	Customer surveys	Response received from customers

TABLE 9.1 (*Continued*)

Potential QOS Measurables

Category	Measurable	Quantifier
	Reliability to acknowledge ship date	Number of orders shipped complete and on time to acknowledge date as a percentage of total orders shipped
	Reliability to request ship date	Number of orders shipped complete and on time to the customer's requested ship date as a percentage of total orders shipped
Manufacturing	Scrap	Percentage of scrap as related to the total order
	Rework	Percentage rework as related to the throughput or the total order
	Run hours per day	Number of hours per day spent running good product
	Throughput per hour	Average good production per hour produced by process
	Setup hours	Hours spent on setup by process
	Downtime	Hours down by reason code by process
	Overall equipment effectiveness	Overall equipment effectiveness by machine
Cost/productivity improvement	Project teams	Quantifiable improvements in areas of team objectives
	Specifications	Annual target $1,000M measured quarterly
Capital investment	Annual capital plan	Rate of return on investment
	Systems	Improved performance in meeting customer requirements
Quality	Process capability	C_{pk} values of all key processes C_{pk} values reported to customer
	Audits	Quantified score from customer #1 Quantified score from customer #2
	Customer quality reports	Summary score from customer #1 Summary score from customer #2
Financial	Inventory turns	Annualized cost of sales/total inventory for current month (sometimes this is measured based on more recent 3 months)
	Sales/employee	Net sales/total employees
	Customer complaint incidents	Number of complaints written in a monthly period—administrative and production

Continued

TABLE 9.1 (*Continued*)

Potential QOS Measurables

Category	Measurable	Quantifier
General	Turnover	Total number of terminations, also voluntary terminations
	Attendance	Total number of days missed Absenteeism documented on daily basis
	Safety	Injury rate is documented and based on incidents per 100 employees

- Brainstorming—to make sure that everyone is participating and contributing to ideas of improvement.
- Nominal group technique—to make sure that legitimate issues are recognized by everyone.
- Pareto chart—to make sure that the significant few are identified as opposed to the trivial many.
- Cause and effect—to make sure that the items of concern have been identified and are really related to the problem at hand.
- Run chart—to make sure consistency is followed.
- Stratification—to make sure that appropriate groups are identified correctly.
- Histogram—to make sure that the data are normally distributed.
- Scatter diagram—to make sure that there is a relationship between variables.
- Control charts—to make sure that the process is in control and stable.
- Capability—to make sure that the product produced meets the requirements of the customer and, more important, to make sure that the machine is capable of producing what the customer wants.
- Force field analysis—to make sure that the problem is clearly understood.

The last key item in the model is to identify and establish correlation and begin to predict external performance. When you have reached this level of analysis in the model, you have indeed identified the key processes as well as the key metrics of those processes. Indeed, you are capable of separating common from assignable cause and you are on your way to achieving your quality improvement targets. Furthermore, because you are able

TABLE 9.2

QOS Implementation

Step	Implementation	Feedback	Model Item
1	Steering committee formation		
2	Steering committee orientation		
3	Customer expectations	Steering committee into continual improvement loop	Customer expectations
4	Strategic goals and objectives	↑	Internal key processes
5	Measurable and processes		Measurable for key internal processes
6	Quantifier/ measurement selection		Trend measurables
7	Action plan formation	Start cascading throughout the organization to various teams	Predict external performance

to predict performance, you can actually adjust the process or processes to new customer expectations. At this point the model repeats itself.

QOS IMPLEMENTATION

Now that we know what QOS is all about, let us examine the implementation process in a given organization. Generally, there are seven steps, which are shown in Table 9.2. Table 9.3 shows the responsibilities for each step.

QOS MEETING SUMMARY

It has been said that "what gets measured it gets done." QOS is no different. To have a successful QOS, regular meetings must take place to review the key processes and measurables and provide effective and applicable feedback. The path to success begins with the current state of your organization—no matter what it is. Generally, there are four steps:

TABLE 9.3

Requirements for Each Step

Step	Implementation	Requirements
1	Steering committee formation	Secure top management involvement
		Explain the QOS process to top management
		Explain the implementation steps
		Select a steering committee
		Explain steering committee responsibilities
		Explain champion responsibilities
		Schedule meetings
2	Steering committee orientation	Review company mission statement and business plan
		Develop steering committee mission
		Identify all customers
3	Customer expectations	List customer expectations
		Group customer expectations
		Rank customer expectations
		Benchmark
4	Strategic goals and objectives	Establish strategic goals and objectives
		Select internal key processes
		Assess team members' knowledge of statistics and variation
		Assess team members' additional training needs
5	Measurable and processes	List potential key processes
		Brainstorm measurable
		Identify measurable for key internal processes
		Assign measurable champions
		Gather existing data on measurable assignment
		Create data management plans
6	Quantifier/measurement selection	Review continual improvement tools for appropriateness
		Consider consensus
		Correlate quantifiers and measurements to customer expectations
		Use problem-solving disciplines
		Track trends of measurables
7	Action plan formation	Present improvement opportunities
		Study opportunities
		Predict external performance
		Identify strengths and weaknesses
		Plan for the future
		Verify and adjust

1. Create a workgroup to assemble a listing of all performance measurements used to operate the organization.
2. Prioritize the most important and create trend reports and goals.
3. Establish the internal measures that correlate directly to customer data.
4. Establish periodic meetings to review the data and the action plans. At the start of the QOS program it is not unusual to have two meetings per week and then eventually one per month.

The flow of the QOS meeting is generally very simple and straightforward. Generally it consists of six items in the following order:

1. Meeting dates
2. List of vital few measurables
3. Prior meeting action items
4. Overview (trend charts for both process and results)
5. Supporting data for each measurable with applicable charts or statistics
6. Lessons learned

During the meeting, the following occur:

- Review prior meeting action items.
- Review trend charts and establish meeting agenda.
- Review adverse trends using problem quantifying tools, make assignments to achieve improvement and/or correction.
- Rotate meeting chairperson to promote team ownership.
- Revalidate measurable quantifiers and quantifying tools on an ongoing basis.
- Issue meeting minute action items.
- Keep an ongoing log of lessons learned and breakthrough experiences.

Example 1

To demonstrate the process and flow of QOS, let us look at an example of generating one. Our example is a baseball QOS.

Step 1. Customer expectations: What do the fans (customers) want of their sport team?
   Answer: Some possibilities are as follows: win the World Series; win American or National League pennant; finish at the top of the

division*; have a winning season. These and many more may be identified through (a) market research, (b) focus groups, (c) satisfaction surveys, (d) field experience, (e) benchmarking, and so on.

Step 2. Internal key processes: Draft; trading of players; pitching/hitting*; defense.

Answer: The key question here is, What is important to the (a) pitcher, (b) batter, (c) coach, and (d) owner? So if we choose the key processes of pitching and batting, the measurables have to reflect those characteristics. A measurable for a pitcher is pitching and for the batter it is hitting. Now we are ready to proceed with the identification of the quantifiers.

Step 3. Measurables for key internal processes: In this example, we decided on pitching and hitting.

Answer: Depending on how we answered step 2, our quantifiers will reflect that selection. For this example we have chosen pitching and hitting as key processes. Therefore, the quantifiers are as follows: for pitching, earned run average [ERA]*, walks, and strike-outs); and for hitting (team average*; home runs; power index; and runs batted in).

Step 4. Trend measurables: Trend with charts the key quantifiers.

Answer: You may do a team ERA chart or a composite comparison chart with last year's performance; you may also do a run chart on the team batting average or a composite comparison with last year's performance. It is very interesting to note that baseball has more statistics for just about anything that you can imagine not only for the team but also for individual players. The reason you do a trend (run) analysis here is to be able to identify any special (sporadic) as opposed to common (chronic) variation.

Step 5. Predict performance: Given the understanding of the key measurables, we can now predict performance.

Answer: We may want to estimate correlation of ERA and attendance. We may want to use scatter diagrams or correlation analysis to find the relationships that interest us.

Note: Items designated with an * have been identified as a key measurable of the process at hand. Of course, these may change depending on what the objective is and who the customer is.

Example 2

Figure 9.2 shows a detailed QOS for a final automotive assembly. The reader will notice that all of the measurables are identified and the goal of 85 percent overall equipment effectiveness (OEE) is reached.

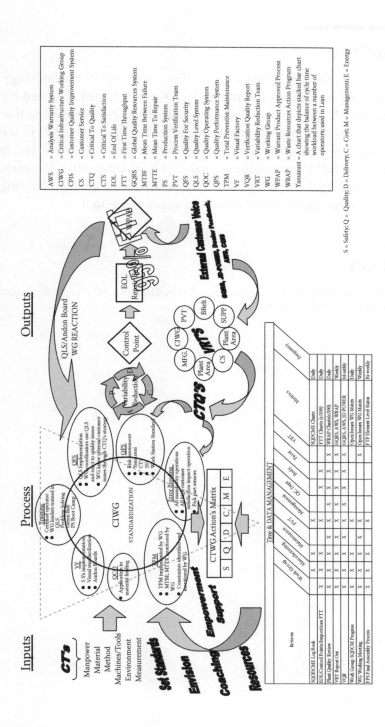

FIGURE 9.2
Specific automotive assembly QOS.

SUMMARY

In this chapter we have given an overview of the QOS methodology. Specifically, we introduced the model for QOS, explained the need for such a methodology, and gave examples using the QOS emphasizing that the Q is not only for the quality department but for all stakeholders in the organization. We closed the chapter with two examples. The first one is based on a baseball team and the second on a final automotive assembly. In the next and final chapter we will address advanced quality planning and its role in the improvement process of any organization.

REFERENCE

Ford Motor Company. (2008). *Quality Operating System*. Dearborn, MI: Corporate Quality Development Center (CQDC).

10

Advanced Quality Planning

In the last chapter we addressed the need for management to have an organizational focus as to how to improve their process in a systematic way using QOS. In this chapter we elaborate the notion that was introduced in the Preface and mentioned in Chapter 2, which was planning. Specifically, we are focusing on this topic once again because in the global environment one of the issues in quality is consistency. To make sure that consistency in requirements and performance within the organization and supply chain exists, some form of advanced quality planning (AQP) must exist.

Formally, AQP is a structured method that defines and establishes the supplier's steps that are necessary to ensure that a product meets customer's requirements (Stamatis 1998). (Note that the term AQP is used interchangeably with APQP—advanced product quality planning. AQP is the generic methodology for all quality planning activities in all industries. APQP is AQP; however, it emphasizes the product orientation of quality. APQP is used specifically in the automotive industry. However, the principle ideas are transferable to other organizations and institutions.)

Once leaders in the world, American companies have lost command of markets to international competitors. Though macroeconomic factors like the exchange rate and trade policies have harmed our ability to compete, a strong case can be made that these problems were chiefly the result of ineffective management practices, as well as the cause of the problems. Among these problems, the lack of planning and vision have been identified by many corporate and governmental personnel as the predominant factors of losing the quality edge to the foreign markets. There are businesses and markets in which U.S. companies no longer compete at all. Those that try to compete find that working harder is not enough and that fundamental changes are necessary.

American companies are fighting back and will have to continue to fight in order to be successful. They are addressing a variety of ways to improve products, services, and return on investment. Some of the more specific approaches include the following:

- More focus on quality and customer satisfaction
- Increased numbers of people involved in quality improvement
- Greater teamwork
- Greater positive coverage in newspapers and magazines
- Introduction of international standards (ISO 9000, ISO 14000; see Appendix A)
- Industry-specific standards (ISO/TS16949, AS9000)
- National awards (Malcolm Baldrige National Award for Quality)
- Intensified training and education
- Use of statistical methods for improvement of quality
- Planning for quality

The results from these changes have been well documented in recent years. However, these changes have been somewhat disjointed, and the competitive position of many of our organizations has not improved substantially. There is a much greater awareness of the urgency for fundamental change in the ways in which organizations operate, as well as a recognition that prevention is better than appraisal systems. One of the realizations is the need for a systematic approach to planning.

A fundamental change that is needed is for quality to be adopted as a business strategy: (a) it must be shown that there is a commitment by the executives and (b) it must be discussed in the board room. This strategy is applicable to all types of organizations, including manufacturing and service companies, schools, hospitals, and government agencies. The aim of this strategy is to enable the organization to produce products and services that will be in demand and to provide a place where people can enjoy their work and take pride in its outcomes.

Quality planning then becomes the means to accomplish the goals and objectives of the organization, such as satisfying the customer, increasing profits or share of the market, growth, better educated citizens, a cleaner environment, lower costs, higher productivity, or increased return on investment. Deming (1986) referred to this as the *quality chain reaction*. Traditionally, increased quality has been thought to come only at the expense of lower productivity and higher cost. This misconception is in part a result of trying to improve quality by inspection or by solving problems rather than by improvements of products and processes.

If quality is to become a business strategy of the twenty-first century, the top managers of the organization must understand quality as a planning strategy and provide leadership for carrying out the strategy. (We introduced this issue in the Preface and touched on it in Chapter 2.) Some important attributes of the planning strategy include the following:

- Provides methods to reach the goals of the organization
- Can be sustained over the long term
- Balances an internal and external focus
- Compatible over different businesses in the organization
- Remains useful despite changes in the marketplace
- Can be understood and practiced by all members of the organization

It has been said many times and in different ways that the difference between planning and no planning may be demonstrated in the bottom line of any organization. The difference, of course, is in the attitude and perseverance of the involved individuals. It is a matter of having a winning and successful attitude in everything we do. From an advanced quality planning perspective, the differences may be articulated in the following way.

- A winner says, "Let's find out."
  - A loser says, "Nobody knows."
- When a winner makes a mistake, he or she says, "I was wrong."
  - When a loser makes a mistake, he or she says, "It wasn't my fault."
- A winner is not nearly as afraid of losing.
  - A loser is secretly afraid of winning.
- A winner works harder than a loser and has more time.
  - A loser is always too busy to do what is necessary.
- A winner goes through a problem.
  - A loser goes around it and never gets past it.
- A winner makes commitments.
  - A loser makes promises. (The attitude here is, to ask for forgiveness rather than permission.)
- A winner says, "I am good, but not as good as I should be."
  - A loser says, "I am not as bad as a lot of people."
- A winner listens.
  - A loser just waits until it is his or her turn to talk.

- A winner respects those who are superior to him or her and tries to learn something from them.
  - A loser resents those who are superior and tries to find chinks in their armor.
- A winner explains.
  - A loser defends or explains away.
- A winner feels responsible for more than his or her job.
  - A loser says, "I only work here."
- A winner says, "There should be a better way to do it."
  - A loser says, "That is the way it is always been done here."
- A winner paces himself or herself.
  - A loser has only two speeds—hysterical and lethargic.
- A winner focuses on advanced planning.
  - A loser focuses on appraisals.

An ever growing number of companies must comply with APQP requirements. Even companies that are not subject to a compliance mandate recognize the APQP process as a product development best practice that improves performance for new product introduction. To implement APQP effectively, companies need to account for a series of key considerations that we have discussed in this book. These considerations will determine the success of the initiative and ultimately the performance of future product launches. The reader is encouraged to read Stamatis (1988) for a detail explanation of the requirements of AQP and Chrysler Corporation, Ford Motor Company, and General Motors (2008) for the specific automotive requirements.

WHY USE ADVANCED QUALITY PLANNING?

Before we address the "why" of planning, we assume that things do go wrong. But why do they go wrong? Obviously, there are many specific answers that address this question. However, all of the answers fall into the following four categories:

- There is never enough time, so things are omitted.
- We have done this this way so that we minimize effort.

- We assume that we know what is been asked for, so we become insufficient.
- We assume that because we finish a project, improvement will indeed follow, so we bypass the improvement steps.

In essence, then, appearance shows that the customer may indeed be satisfied, but a product, service, or the process is not improved at all. This is precisely why it is imperative for organizations to look at quality planning as a totally integrated activity that involves the entire organization. The expectation of the organization must be to accept changes in the operations by employing cross-functional and multidisciplinary teams to exceed customer desires—not just meet requirements. A quality plan includes, but is not limited to

- A team to manage the plan.
- Timing to monitor progress.
- Procedures to define operating policies.
- Standards to clarify requirements.
- Controls to stay on course.
- Data and feedback to verify and to provide direction.
- An action plan to initiate change.

AQP, then, is a methodology that allows for maximum quality in the workplace by planning and documenting the process of improvement (Figure 10.1). The figure visually shows the flow of continual improvement by adopting the plan–do–check–act (PDCA) model. In evaluating (looking) at this model, one can also see how it is applicable in any situation where a concept is generated and, through appropriate and applicable approvals, produced. In addition, because of the C and A stages, one is able to not only check the product and/or service but to use the model for both corrective action and feedback. In this sense, AQP is the essential discipline that offers both the customer and the supplier a systematic approach to quality planning, defect prevention, and continual improvement. Some specific uses are the following:

- In the auto industry in particular, the demand is so high that Chrysler, Ford, and General Motors have developed a standardized approach to AQP. That standardized approach is a requirement for ISO/TS16949 certification. In addition, each company has its own

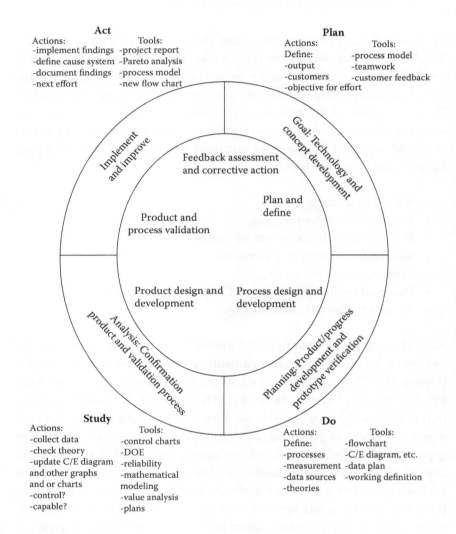

Act

Actions:
-implement findings
-define cause system
-document findings
-next effort

Tools:
-project report
-Pareto analysis
-process model
-new flow chart

Plan

Actions:
Define:
-output
-customers
-objective for effort

Tools:
-process model
-teamwork
-customer feedback

Implement and improve

Goal: Technology and concept development

Feedback assessment and corrective action

Plan and define

Product and process validation

Product design and development

Process design and development

Analysis: Confirmation product and validation process

Planning: Product/progress development and prototype verification

Study

Actions:
-collect data
-check theory
-update C/E diagram and other graphs and or charts
-control?
-capable?

Tools:
-control charts
-DOE
-reliability
-mathematical modeling
-value analysis
-plans

Do

Actions:
Define:
-processes
-measurement
-data sources
-theories

Tools:
-flowchart
-C/E diagram, etc.
-data plan
-working definition

FIGURE 10.1
Overview of an AQP model.

way of measuring success in the implementation and reporting phase of AQP tasks.

- Auto suppliers are expected to demonstrate the ability to participate in early design activities from concept through prototype and on to production.
- Quality planning is initiated as early as possible, well before print release.
- Planning for quality is needed particularly when a company's management establishes a policy of prevention as opposed to detection.
- When you use advanced quality planning, you provide for the organization and resources needed to accomplish the quality improvement task.
- Early planning prevents waste (scrap, rework, and repair), identifies required engineering changes, improves timing for new product introduction, and lowers costs.

WHEN DO WE USE ADVANCED QUALITY PLANNING AS AN EFFECTIVE METHODOLOGY?

We use AQP as an effective methodology when we need to meet or exceed our expectations in the following situations:

- During the development of new processes and products
- Prior to changes in processes and products
- When reacting to processes or products with reported quality concerns
- Before tooling is transferred to new producers or new plants
- Prior to process or product changes affecting product safety or compliance to regulations

In the case of an automotive supplier, the supplier is to maintain evidence of the use of defect prevention techniques prior to production launch. The defect prevention methods used are to be implemented as soon as possible in the new product development cycle.

It follows, then, that the basic requirements for an appropriate and complete AQP are as follows:

- A team approach
- Systematic development of products/services and processes
- Reduction in variation so that the customer the customer may be delighted (this must be done even before the customer requests improvement of any kind)
- Development of a control plan

As AQP is continuously used in a given organization, the obvious need for its implementation becomes stronger and stronger. That need may be demonstrated through the following:

- Minimizing the present level of things gone wrong.
- Yielding a methodology that integrates customer and supplier development activities, as well as concerns.
- Exceeding present reliability/durability levels to surpass the competition's and customer's expectations.
- Reinforcing the integration of quality tools with the latest management techniques for total improvement.
- Exceeding the limits set for cycle time and delivery time.
- Developing new and improving existing methods of communicating the results of quality processes for a positive impact throughout the organization.

HOW DO WE MAKE AQP WORK?

Just as with any other program, there are no guarantees for making AQP work. However, there are three basic characteristics that are essential and must be adhered to in order for AQP to work:

1. Activities must be measured based on the who, what, where, and when guidelines.
2. Activities must be tracked based on shared information (how and why), as well as work schedules and objectives.
3. Activities must be focused on the superordinate goal of quality–cost–delivery, using information and consensus to improve quality.

As long as our focus is on the triad of quality–cost–delivery, AQP can produce positive results. After all, we all need to reduce costs while we increase quality and reduce lead time. That is the focus of an AQP program, and the more we understand it, the more likely we are going to have a workable plan.

ARE THERE PITFALLS IN PLANNING?

Just like everything else, planning has pitfalls. However, if one considers the alternatives, there is no doubt that planning will win out. Perhaps one of the greatest pitfalls in planning is the lack of support by management and a hostile climate for its practice. So, the question is not really whether or not are there any pitfalls but why such support is quite often withheld and why such climates occur in organizations that claim to be quality oriented.

We suggest that perhaps some specific pitfalls in any planning environment may have to do with commitment, time allocation, objective interpretations, a tendency toward conservatism, and an obsession with control. All of these elements breed a climate of conformity and inflexibility that favors incremental changes for the short term but ignores the potential of large changes in the long run. Of these, the most misunderstood element is commitment.

The assumption is that with the support of management, all will be well. This assumption is based on the axiom of F. Taylor at the turn of the twentieth century, which is "There is one best way (2010)." Planning is assumed to generate the one best way not only to formulate but to implement a particular idea, product, and so on. Sometimes, this notion is not correct and thereby unnecessary problems arise. In our modern world we must be prepared to evaluate several alternatives of equal value.

As a consequence, the issue is not simply whether management is committed to planning. Mintzberg (1994) observed that it also is (a) whether planning is committed to management; (b) whether commitment to planning engenders commitment to the process of strategy making, to the strategies that result from that process, and ultimately to the taking of effective actions by the organization; and (c) whether the very nature of planning actually fosters managerial commitment to itself.

Another pitfall of equal value is the cultural attitude of "fighting fires." In most organizations, we reward problem solvers rather than planners. As a consequence, in most organizations the emphasis is on low-risk firefighting, when in fact it should be on planning a course of action that will be realistic, productive, and effective. Planning may be tedious in the early stages of conceptual design, but it is certainly less expensive and much more effective than corrective action in the implementation stage of any product or service development.

DO WE REALLY NEED ANOTHER QUALITATIVE TOOL TO GAUGE QUALITY?

Qualitative studies have been viewed as a way to measure competitiveness rather than a component to any quantitative analysis. Though quantitative methods are excellent ways to address the who, what, when, and where, qualitative study focuses on the why. It is in this why that the focus of advanced quality planning contributes the most results, especially in the exploratory feasibility phase of our projects.

So, the answer to the question is a categorical "yes," because the aim of qualitative study is to understand rather than to measure; it is used to increase knowledge, clarify issues, define problems, formulate hypotheses, and generate ideas. Using qualitative methodology in AQP endeavors will indeed lead to a more holistic, empathetic customer portrait that can be achieved through quantitative study, which, in turn, can lead to enlightened engineering and production decisions as well as advertisement campaigns.

HOW DO WE USE THE QUALITATIVE METHODOLOGY IN AN ADVANCED QUALITY PLANNING SETTING?

Because this book is focused on addressing ideas that are inherently important for improvement in the twenty-first century rather than a treatise of each of the items identified, allow me to summarize the methodology in seven steps:

1. *Begin with the end in mind.* This may be obvious; however, this is how most goals are achieved. This is the stage where the experimenter determines how the study results will be implemented: What courses of action can the customer take, and how will they be influenced by

the study results? Clearly, understanding outcomes defines the study problem and report structure. To ensure implementation, determine what the report should look like and what it should contain.

2. *Determine what's important.* All resources are limited and therefore we cannot do everything. However, we can do the most important things. We must learn to use the Pareto principle (vital few as opposed to the trivial many). To identify what is important, we have many methods, including asking about advantages and disadvantages, benefits desired, likes and dislikes, importance ratings, preference regression, key driver analysis, conjoint and discrete choice analysis, force field analysis, value analysis, and so many others. The focus of all of these approaches is to improve performance in areas in which a competitor is ahead or in areas in which your organization is determined to hold the lead in a particular product and/or service.

3. *Use segmentation strategies.* Not everyone wants the same thing. Learn to segment markets for specific products and/or services that deliver value to your customer. By segmenting based on wants, the engineering and/or product development can provide action-oriented recommendations for specific markets and therefore contribute to customer satisfaction.

4. *Use action standards.* To be successful, standards must be used with diagnostics. These standards must be defined at the outset. When the results come in, there will be an identified action to be taken, even if it is to do nothing. List the possible ways in which the results can come out and the corresponding actions that could be taken for each. If you cannot list actions, then you have not designed an actionable study. Design it again.

5. *Develop optimals.* Everyone wants to be the best. The problem with this statement is that there is only room for one best. All other choices are second best. When an organization focuses on being the best in everything, that organization is asking for a failure. No one can be the best in everything and sustain it. What we can do is focus on the optimal combination of choices. By doing so, we usually have a usable recommendation based on a course of action that is reasonable and within the constraints of the organization.

6. *Give grasp-at-a-glance results.* The focus of any study is to turn people into numbers (wants into requirements), numbers into a story (requirements into specifications), and that story into action (specifications into products or services). But the story must be easy to

understand. The results must be clear and well organized, so that they and their implications can be grasped at a glance.

7. *Recommend clearly.* Once you have a basis for an action, recommend that action clearly. An analogy may prove the point. You do not want a doctor to order tests and then hand you the laboratory report. You want to be told what is wrong and how to fix it. From an advanced quality planning perspective, we want the same. That is, we want to know where the bottlenecks are, what kind of problems we will encounter, and how we will overcome them for a successful delivery.

SUMMARY

In this closing chapter we have discussed AQP. Specifically, we defined what AQP is and why it is necessary and provided a general model to follow. We also emphasized the need for any organization to pursue planning as opposed to appraisal approaches because planning is much more economical than appraising systems. Planning reduces cost, improves quality, and optimizes scheduling.

REFERENCES

Chrysler Corporation, Ford Motor Company, and General Motors. (2008). *Advanced product quality planning and control plan*, 2nd ed. Southfield, MI: Automotive Industry Action Group.

Deming, W. E. (1986). *Out of the crisis.* Cambridge, MA: Massachusetts Institute of Technology, Center for Advanced Engineering Study.

Mintzberg, H. (1994). *The rise and fall of strategic planning.* New York: Free Press.

Stamatis, D. H. (1998). *Advanced quality planning.* New York: Quality Resources.

Taylor, F. W. (2010). *The Principles of Scientific Management.* Sioux Falls, SD: Ezreads Publications, LLC.

Epilogue

Now more than ever we need to reevaluate our values, standards, behaviors, and attitudes about our future as a country. The change is coming upon us as an avalanche down a mountain full of snow. Unless we take cover, we are indeed in danger. It is time to act with boldness and tell the truth. Part of the boldness has to be in the reevaluation and the truthfulness of the situation at hand. Part of the evaluation has to do with quality. How we view quality, how we analyze it, and how we report it have become major concerns and a topic of conversation in many circles in both manufacturing and nonmanufacturing. Obviously, the last one (reporting) is very fickle and gives a variety of answers because the selection of the characteristics will determine the reporting methodology as it is attested to by the cost of quality and the Sarbanes–Oxley Act that many organizations are grappling with. The other two (viewing and analyzing), however, are just as important but for some reason we seem to superficially pretend to understand the principles and apply them accordingly.

In this book we have examined some of these principles and proposed a true commitment to true quality for significant improvement. So, here we go.

We must come to grips—all of us—with the fact that something drastic has to happen. All around us things are changing. Some of the changes are for the better, but a lot for the worse. Outsourcing is rampant and our manufacturing base is shrinking. We see unemployment for high-paying jobs diminishing and costs for doing business are increasing in an unprecedented manner. Some of the causes are unfair trade policies, litigation, health care, quality, innovation, and general wages.

History tells us something about this degradation. All civilizations throughout human history have lasted for about 200 years or so as a mighty power and then they begin to decline. We saw this decline in the Egyptian, Babylonian, Persian, Greek, Roman, British, and other empires. What is interesting about the decline in each one of these empires is that the source of the decline was the same. It turns out that the central cause is moral decay. This in turn precipitated a military decline (in the form of a voluntary and mercenary army); economic decline (in the form of taxation and imports of goods from conquered lands); legal and social decline (in the form of "everything goes" as long as the citizenship wants it and makes them feel good); political decline (in the form of public corruption, the inability to face and solve serious issues of the day, and the size of bureaucracy).

In today's world, we do see the similarities, at least in the United States, even though we are still considered to be a superpower. But the question has become "For how long?" I submit that perhaps we may be able to change the course of history if we at least face up to some realities. I cannot help but think of the dynamic, motivational speech that General George S. Patton gives in the opening scene of the movie *Patton*[1]:

Now, I want you to remember ... that no bastard ever won a war ... by dying for his country. He won it ... by making the other poor dumb bastard die for his country. Men ... all this stuff you've heard about America not wanting to fight ... wanting to stay out of the war ... is a lot of horse dung. Americans ... traditionally love to fight. All real Americans love the sting of battle. When you were kids ... you all admired the champion marble shooter ... the fastest runner, big-league ball players, the toughest boxers. Americans love a winner ... and will not tolerate a loser. Americans play to win all the time. I wouldn't give a hoot in hell for a man who lost and laughed. That's why Americans have never lost and will never lose a war ... because the very thought of losing ... is hateful to Americans. Now ... an army is a team. It lives, eats, sleeps, fights as a team. This individuality stuff is a bunch of crap. The billion bastards who wrote that stuff about individuality ... for the *Saturday Evening Post* ... don't know anything more about real battle than they do about fornicating. Now we have the finest food and equipment ... the best spirit ... and the best men in the world. You know ... by God, I actually pity those poor bastards we're going up against. By God, I do. We're not just going to shoot the bastards ... we're going to cut out their living guts ... and use them to grease the treads of our tanks. We're going to murder those lousy Hun bastards by the bushel. Now ... some of you boys ... I know are wondering ... whether or not you'll chicken out under fire. Don't worry about it. I can assure you ... that you will all do your duty. The Nazis ... are the enemy. Wade into them! Spill their blood! Shoot them in the belly! When you put your hand ... into a bunch of goo ... that a moment before was your best friend's face ... you'll know what to do. There's another thing I want you to remember. I don't want to get any messages saying we are "holding our position." We're not "holding" anything. Let the Hun do that. We're advancing constantly. We're not interested in holding on to anything ... except the enemy. We're going to hold on to him by the nose and kick him in the ass. We're going to kick the hell out of him all the time ... and we're going to go through him like crap through a goose! Now ... there's one thing ... that you men will be able to say when you get back home. And you may thank God for it. Thirty years from now when you're sitting around your fireside ... with your grandson on your knee ... and he asks you: "What did you do in the great World War II?" You won't have to say: "Well ... I shoveled shit in Louisiana." All right, now, you sons of bitches ... you know how I feel. I will be proud ... to lead you wonderful guys into battle anytime ... anywhere. That's all.

What an inspiring speech! However, in today's environment not only it is not politically correct to speak this way, but our leaders are trying to find excuses for losing or trying to justify their actions for defeat. Worse yet, losers are rewarded by either government bailouts or golden parachutes. Times have changed indeed!

Why mention Patton's encouraging words in a quality discussion? Because his oratory delight (in form, delivery, and content) is appropriate and applicable in our times. One can write a book about the parallels of what he said and how we are all behaving today. So let us look at some of the parallels.

WE ARE INDEED IN AN ECONOMIC WAR THROUGH GLOBALIZATION

Whereas Patton was fighting the Nazis, we are fighting in terms of commerce everywhere. The problem, however, is that we are not fighting with equal trade policies. It is unfortunate indeed that if someone begins to talk about these inequalities they are labeled either racists, close-minded, protectionists—the list goes on. In the meantime, the erosion continues at the expense of the United States. The fallacy of this economic world war is that if we do not open our markets we will not be able to compete. On the other hand, we are open to foreign markets but we give the store away in the process. The end result is record deficits. An example will suffice here: I remember in my first operations management class (in the late 1960s) the professor told us that the key to world dominance ever since the industrial revolution is industrial dominance, pure and simple. That was the time when manufacturing was indeed thriving in the United States. We had entire cities dedicated to particular products. For example, Detroit, automobiles; Akron, rubber; Pittsburgh, steel; New York, financial services; Boston, shoes and insurance; and so on. Now we see that these same cities are fighting for survival on many fronts.

On the other hand, China, the so-called the land of opportunity, focuses on the city specialization of our past. So we see entire cities committed to producing socks, shoes, automobiles, and so on. China's growth is hard to calculate, but estimates are approximately 10 percent per year. But we forget that that growth is a direct hit against us. For the time being we are gaining on price differentiation, but what will happen to us 30 years from now? If we plan to win, why do we give the competitors the tools to defeat us? After all, let us not forget the words of Stalin when he said "The capitalists will sell us the rope with which we will hang them."

Mexico is another example. We force our manufacturers to deal with taxation policies, standards, and requirements so that in the short term it is cost-effective to move elsewhere, and at the same time we complain about unemployment and foreign trade imbalances. It is about time to equalize the field of trade.

AMERICANS LIKE TO FIGHT. THEY HATE THE IDEA OF LOSING. THEY PLAY TO WIN

One of the American characteristics is competition—open and fair competition. It is through competition that we improve our products in the sense of better prices, better quality, and, above all, better efficiency. It has been one of the trademarks of American ingenuity; that is, to find a better way to do things and drive the competitors out of the market.

Unfortunately, in the last 15–20 years America has been exposed to some unfair practices and, in conjunction with the myopia of short-term profits in terms of quarterly earnings, we all have suffered. Quality has indeed played a role in this equation; however, if we look closely at what is happening, especially in the automotive market, we see that the gap is closing drastically, but foreign producers still outsell domestic producers in record numbers. Yes, Americans like to fight and win, but they cannot win with unbalanced rules that favor foreign competition. They cannot win when the standards of operations are so far apart that foreign manufacturers operate with practically no constraints.

Americans hate the idea of losing in any endeavor. However, if executives and politicians do not support the employees, there will be no battle to win. Executives focus on the strategy of the organization and they do have the ability to do something about it. In fact, if executives begin by receiving payment based on performance, things will change drastically. Currently, there is practically no incentive for executives to make sure that the organization will benefit because his or her reward (in most cases) is a set contractual agreement. So we see companies going out of business, but the executives still make millions of dollars.

Politicians, on the other hand, have an equal responsibility in this battle. They must be watchful and be aware of treaties, memorandums, and trade agreements that do not second-guess the productivity and profitability of American companies. Yes, we can and should help others, but not so much that we begin to affect our own economy.

To be a winner, all of us must play the appropriate role. We must be a team—a team with a vision; with a doable goal; with conviction; with enthusiasm. A team with the desire to win.

WE HAVE THE BEST

Through no fault of the workers, our manufacturing and other industrial sectors are suffering because of the inability of their leaders to forecast and plan for the needs of the future. Our workers do have high qualifications, and they deserve the best from their leaders. You cannot blame the workers when the system is defined, directed, and implemented through the discretion of management.

No one will deny that we do have very qualified workers in many sectors of our economy. No one will deny that we have appropriate and applicable systems to collect and evaluate information that will make us understand the process at hand. No one will deny that our training methodologies for the latest innovations are available. On the other hand, no one will deny that all of these items are not consistently implemented and quite often are seen as chores rather that improvement items. The fact that we keep on changing quite often for no reason other than that we have to have the latest contributes to problems. These problems are not employee problems; rather, they belong to management.

TO WIN WE MUST BE COMMITTED.
WE MUST HAVE AN ENEMY TO FOCUS

One of the memorable moments in Patton's story was when his troops were marching in bad weather on muddy roads. There was the general— in the middle of the soldiers, meddling with them, encouraging them, and fighting the mud, the snow, and the elements of nature. The soldiers understood his commitment. They felt his conviction. They sensed his faith in them. They knew that their enemy (Nazis) was his enemy. By contrast, we see that the executives of today's organizations focus on short-term profits rather than the long-term effects of their goals and decisions. For example, in the mid-1960s, GM had a market share of about 48.3 percent, Ford had about 25.2 percent, Chrysler had 14.1 percent, Other US and imports had 7.7 percent, American Motors had 4.2 percent, and Studebaker had .5 percent (Earl, 2005). In 2010 the market had changed drastically to the point that the percentages for the automotive companies were dramatically different as some automobiles had disappeared from the market and new ones had entered the same market. So we see GM to be at 17.6 percent, Ford at 16.3 percent, Chrysler at 9.7, Nissan at 7.5, Toyota at 15.6, Hyundai/Kia at 8.9, Honda at 11 percent, and all others 13.4 percent (Schepp, 2010). In the last 10 years alone the changes for the major American companies have been demonstrating a drop of about 10 percent for each company. In fact, GM declared bankruptcy in 2009, and Chrysler declared bankruptcy in 2007 and was bought by Fiat in 2009.Yet, the executives have consistently received millions in salary and many more millions in stock options. And the workers? They are threatened with wage concessions, plant closings, health care (insurance) and pension reductions, and so on. In contrast, Japanese companies not only have been increasing their market share but have also increased their profitability with the methodologies and approaches to manufacturing that we taught them.

Their commitment seems to be conquer their enemy—the United States market. We, on the other hand, are committed to increasing the profits of our executives and showing profitability to the stockholders at the end of the quarter. The enemy is all of the forces that have ganged up against the United States to make our standard of living much lower and somewhat comparable with that of the rest of the world. To fight the enemy and its objective, all of us must be mobilized to make sure that our economy is the best it can be and that it continues to be for a long time. That is an issue that executives and politicians must realize, and they should implement solutions for revitalizing as well as gaining the faith of their own employees and constituents. We cannot continue on the current path with an attitude of "got you" and blame someone else. We cannot talk about quality as being important and yet have recalls of our products in record numbers for simple failures.

ALWAYS MOVE FORWARD. DO NOT RETREAT; LET YOUR ENEMY DO THAT

Patton's remarks are again timeless. Yet executives and politicians alike seem to have given up. Retreat should be the last thing we do. Instead, we should echo the 300 Spartan's proclamation ("Come and get it") to the Persians in the Battle of Thermopele, or the response of the few in defending the Alamo, or the response of the American Army to the Nazi command with a simple "nuts," when they were asked to surrender. If the commitment is strong and the goal crystal clear, there is nothing to retreat from or give up. Rather, it becomes a self-fulfilling prophesy for success. For example, If everyone says that there is no profit in building small cars and this is repeated many times, people begin to believe it. On the other hand, Japanese companies are laughing all the way to the bank. If one says that style is the answer, Japanese and German companies are laughing all the way to the bank. If safety were the issue, Volvo would be the number one automobile in the world. Instead, Volvo was bought by Ford and then resold. However, if we examine the commitment and specificity of the Japanese long-term goals and objectives, we see that they have come a long way from the 1960s. How? By focusing on quality; that is, year after year, a consistent improvement (reduction) in the number of vehicle defects. It took them a long time to catch up. But now they are the ones who lead. American companies, even today, dismiss the billions of dollars that they spend in rework and warranty (for some companies over $5.2 billion per year) as part of doing business. As a result, the customer suffers and business profitability decreases, even though executive salaries remain very high for what they deliver.

HAVE LONG-RANGE SCOPE

The mess we are in did not occur overnight. As a consequence, our turn-around will take some time. How are we going to accomplish this? First we must mean what we say about quality. If quality is indeed a primary concern, we cannot afford to find excuses to produce more and then recall the items. We have the methodologies for improvement. We commit millions of dollars to training employees. What is missing is the commitment to excel, the inherent enthusiasm of winning and being successful. We have indeed become complacent. We are fearful of being number one; we are scared to death that we are going to ruffle feathers in our pursuit of greatness.

The long-range scope must start right way. It is an issue of appropriate planning. It is an issue of reorganizing our thoughts and goals as well as perspective. We must believe that we can do it. After all, despite our scientific achievements, Western society has allowed itself to be held hostage by doomsayers and politicians eager to gain advantage. We live in a country where citizens live longer, healthier lives than ever before and pollution levels continue to decline despite greater economic activity, yet we have convinced ourselves that the end is near. It probably is but only because we have chosen to chase false gods who will save us from an impending disaster that will never come.

BE COMMITTED TO YOUR PEOPLE. SHOW YOUR LEADERSHIP AND PEOPLE WILL FOLLOW

There is no question that the United States has today a huge head start on China and the rest of the world. We are indeed the number one market the world wants to sell their products in. Our country has an enormous advantage in military power, gross national income, and many other superlatives. However, if we rest on these past laurels, we will certainly fall behind and other countries will become leaders.

To take advantage of our power (military, economic, and social), we must recognize that the government must play a role in this reconfiguration (adjustment), and executives must begin leading their organizations in terms of benefits not only for themselves and their stockholders but their employees. As for the government's role, the issue is trade. They must think of ways to make trade fair through treaties or global agreements that put everyone on the same level. Executives must understand that the battle is going to be won only if everyone participates—and that includes employees. Leadership is needed to guide us in this Herculean task through the labyrinth of convoluted practices, behaviors, attitudes, and specific contracts. However, above all, to gain the cooperation of all employees, leaders must have integrity, gain their trust, be truthful, and work with them for positive results.

One of the issues in this chain includes the recognition that perhaps our expectations may be changed. For example, incentives to buy a certain product/service may have to be reevaluated because they are not effective.

Leaders must lead by example. Words are not good enough. Slogans and euphemisms are no longer effective. Quotas for productivity are nothing more than an easy way to say "We are trying." All of us who have been involved with quality for some time know that quotas do not work and, most important, frustrate the employees who produce not for quality but rather for production.

Employees see this and they translate it as "If they do not care, why should we?" Employees need a leader who is committed to what will benefit the company and the employees. For example, GM's situation is critical on many fronts; that is, sales, market share, retiree pensions, health care, trade policies, and so on. The answers to these questions are not dependent upon the employees, because they have no control over them. They are, on the other hand, a very integral part of leadership. It is how the leaders react and decide and then sell it to the employees that matters. GM obviously is not unique. The crisis in the United States is bigger than any one company or even any one industry. It represents a crossroads for our economy and our country. However, this presents us with an opportunity to choose for America and its families a future that we can all live with based on our expectations of a free society.

ENDNOTE

1. *Patton* won seven Oscars in 1971, including those for Best Picture, Best Director, Best Original Screenplay, and Best Actor. The film was directed by Franklin J. Schaffner, from a script by Francis Ford Coppola and Edmund H. North, and starred George C. Scott in the title role. It was based on *Patton: Ordeal and Triumph* by Ladislas Farago and *A Soldier's Story* by General Omar Bradley.

REFERENCES

Earl, H. (2005). http://www.carofthecentury.com/answer_to_gm%27s_market_share_plunge.htm.

Schepp, D. (September 23, 2010). "GM's Market Share Hovers Near Record Low." *Daily Finance*. http://www.dailyfinance.com/2010/09/23/gms-market-share-hovers-near-record-low/.

Selected Bibliography

Abbate, F. (2007). "Why ergonomics is only half the answer." *Office Solutions* January, 37–39.

Alster, N. (2008). "Customer disservice?" *CFO* Fall, 20–27.

Anderson, B. (2008). "A framework for business ethics." *Quality Progress* March, 22–29.

Anderson, J. A. (1984). "Regression and ordered categorical variables." *Journal of the Royal Statistical Society Series B* 46: 1–30.

Anderson, T. (2006). "Listen to the text." *Quirk's Marketing Research Review* October, 52–56.

Arian, N. (2008). "Teams: It all ties together." *Quality Progress* May, 20–27.

Aust, S. (2000). "Ensuring successful simulation modeling project." *Quality Digest* November, 28–31.

Awad, A. and M. Usmen. (2006). "12 Components of integrated construction project management." *Technology Century* August–September, 39–43.

Bahlis, J. (2008). "Blueprint for planning learning." *T&D* March, 64–67.

Bailey, W. M. (2006). "Evaluating a market using P-E gap analysis." *Quirk's Marketing Research Review* June, 22–26.

Bala, S. (2008). "A practical approach to Lean Six Sigma." *Quality Digest* July, 30–34.

Balestracci, D. (2003). "Handling the human side of change." *Quality Progress* November, 38–45.

Bangert, M. (2007). "The power of document control." *Quality* February, 40–45.

Banton, M. (1997). *Ethnic and racial consciousness*. London: Longman.

Barrows, M. (2005). "Use distribution analysis to understand your data source." *Quality Progress* December, 50–56.

Beffa, X.-L. and X. Ragot. (2008). "The fall of a financial model." *Financial Times* 22 February.

Bell, A. (2006). "Avoiding the leadership trailspin." *T&D* November, 66–67.

Benitez, Y., L. Forrester, C. Hurst, and D. Turpin. (2007). "Hospital reduces medication errors using DMAIC and QFD." *Quality Progress* January, 38–45.

Berry, L. L. and L. P. Carbone. (2007). "Build loyalty through experience management." *Quality Progress* September, 26–32.

Bester, Y. (1993). "Net-value productivity: Rethinking the cost of quality approach." *Quality Management Journal* October, 71–76.

Betof, E. (2007). "The rock and the fish: Teachable points of view for leadership." *T&D* March, 48–53.

Bingham, T. and P. Galagan. (2007). "Finding the right talent for critical jobs." *T&D* February, 30–36.

Bingham, T. and T. Jeary. (2007). "Communicating the value of learning." *T&D* May, 80–84.

Blumberg, R. (2006). "Time well spent." *Quirk's Marketing Research Review* October, 34–39.

Boebinge, D. (2006). "You think you are not a project manager." *Technology Century* August–September, 29–31.

Boehm, T. C. and J. M. Ulmer. (2008). "Product liability: Beyond loss control—An argument for quality assurance." *Quality Management Journal* 15(2): 7–18.

Borawski, P. E., G. L. Duffy, T. Foster Jr., H. M. Guttman, L. B. Hare, D. Okes, G. C. Payne, J. J. Rooney, J. West, and R. Westcott. (2007). "Quality basics." *Quality Progress* June, 25–36.

Brandt, R. (2000). "Loyalty really isn't all that simple, restrictive." *Marketing News* 14 August.

Brennan, J. (2003). "Reality check: The new learner needs." *T&D* May, 23–25.

Buck, T. (2006). "Europe regulators outpunch U.S." *Financial Times* 3 November.

Bullington, K. (2009). "Driven to succeed." *Quality Progress* February, 46–51.

Byham, W. C. (2002). "14 Leadership traps." *T&D* March, 56–63.

Byham, W. C. (2008). "Flexible phase-out." *T&D* April, 34–37.

Cacioppo, K. (2000). "Measuring and managing customer satisfaction." *Quality Digest* September, 23–24.

Cameron, K. and M. Lavine. (2007). *Making the impossible possible.* San Francisco: Berret-Koehler Publishers.

Carlson, S. (2006). "In search of the sustainable campus." *The Chronical of Higher Education* 20 October.

Chowdhury, S. (2000). "Changing management styles put their mark on industry." *Quality Progress* May, 61–66.

Clutterbuck, D. (2002). "How do teams learn." *T&D* March, 67–69.

Cohen, E. (2007). *Leadership without borders.* New York: John Wiley & Sons.

Coldwell, K. (2008). "Managing outcomes in a Lean enterprise." *Quality* November, 40–41.

Cole, R. E. (1993). "Learning from learning theory: Implications for quality improvement of turnover, use of contingent workers, and job rotation politics." *Quality Management Journal* October, 9–25.

Collins, K. F. (2007). "Applying the Toyota Production System to a healthcare organization: A case study on a rural community healthcare provider." *Quality Management Journal* 14(4): 41–52.

Conklin, J. D. (2006). "Measurement system analysis for attribute measuring processes." *Quality Progress* March, 50–53.

Cottrell, D. (2002). *Monday morning leadership.* Dallas, TX: CornerStone Leadership Institute.

Cox, T. (2008). "Six Sigma: Map quest." *Quality Progress* May, 44–52.

Crago, M. G. (2002). "Meeting patient expectations." *Quality Progress* September, 41–44.

Crawford-Mason, C. (2002). "Deming and me." *Quality Progress* September, 45–49.

Crosby, P. B. (2005). "Crosby's 14 steps to improvement." *Quality Progress* December, 60–66.

Dalgleish, S. (2007). "Quality professionals need to shift focus to short-term results." *Quality* February, 14.

Davenport, R. (2005). "Faith at work." *T&D* December, 22–29.

De Feo, J. A. (2005). "A road for change." *Quality Progress* January, 24–30.

De Feo, J. A. (2007). "The battle for quality." *Quality Digest* July, 44–47.

Dearing, J. (2007). "ISO 9001: Could it be better?" *Quality Progress* February, 23–27.

Derven, M. (2008). "Management onboarding." *T&D* April, 49–52.

Dew, J. (2003). "The seven deadly sins of quality management." *Quality Progress* September, 59–64.

DiOrio, L. J. (2003). "The proven way: The system engineering approach." *Quality Progress* September, 46–50.

Dodd, P. (2005). *The 25 best time management tools and techniques.* Denver, CO: Peak Performance Press.

Doggett, A. M. (2005). "Root cause analysis: A framework for tool selection." *Quality Management Journal* 12(4): 34–45.

Dressler, L. (2007). *Consensus through conversation*. San Francisco: Berret-Koehler Publishers.

Economist. (September 24, 2008). "And then there were none." http://www.economist.com/node/12294688.

Ericson, J. (2006). "Lean inspection through supplier partnership." *Quality Progress* November, 36–47.

Erker, S. (2007). "What does your hiring process say about you?" *T&D* May, 66–71.

Evans, J. M. (2006). "Look for trouble." *Quality Progress* December, 56–62.

Fazi, C. (2007). "SOPs relay knowledge." *Quality* February, 46–51.

Feigenbaum, A. V. (2008). "Cost of quality: Raising the bar." *Quality Progress* July, 22–27.

Feingold, J. and B. Miller. (2006). "Leading a Lean initiative from the ranks." *Quality Digest* December, 40–44.

Fine, E. S. (1998). "Are you listening to your customers?" *Quality Progress* January, 120.

Fisher, F. (2007). "Clean power is key to the future." *The Age* 10 May.

Fontenot, G., L. Henke, and K. Carson. (2005). "Take action on customer satisfaction." *Quality Progress* July, 40–47.

Foster, S. T. (2006). "One size does not fit all." *Quality Progress* July, 54–61.

Foster, S. T. (2007). "Does Six Sigma improve performance?" *Quality Management Journal* 14(4): 7–20.

Francis, L. (2007). "Mentoring makeover." *T&D* July, 52–57.

Fredericks, J. O. and J. M. Salter II. (1998). "What does your customer really want." *Quality Progress* January, 63–68.

Friedman, D. and D. Giber. (2007). "Making or breaking outsourcing success." *T&D* September, 55–57.

Friedman, L. (2006). "Are you losing potential new hires at hello?" *T&D* November, 25–27.

Fuchs, E. (1993). "Total quality management from the future: Practices and paradigms." *Quality Management Journal* October, 26–34.

Gale, B. T. (1994). *Managing customer value*. New York: The Free Press.

Gardenswartz, L., J. Cherbosque, and A. Rowe. (2009). "Coaching teams for emotional intelligence in your diverse workplace." *T&D* February, 46–49.

Gasper, B. (2009). "Whose need is it anyway." *Quirk's Marketing Research Review* January, 60–64.

Gay, D. L. and T. J. LaBonte. (2003). "Demystifying performance: A roadmap." *T&D* May, 64–75.

Gebauer, J. (2006). "Workforce engagement." *T&D* November, 28–30.

Goedeking, P. (2007). "The challenge in China" *Automotive Design and Production* February, 18.

Goldstein, S. (2007). "Using statistics to improve satisfaction." *Quality Progress* March, 28–34.

Goleman, D., R. Boyatzis, and A. McKee. (2006). *Primal leadership*. Boston: Harvard Business School Press.

Goodden, R. L. (2008). "Product safety: Better safe than sorry." *Quality Progress* May, 28–35.

Goodman, J. and C. D. Collier. (2007). "Deliver great service by listening and adapting." *Quality Progress* March, 22–27.

Gordon, D. (2009). "Risk and quality management." *Quality Progress* January, 60–61.

Govindarajan, V. and C. Trimble. (2010). *The other side of innovation: Solving the execution challenge*. Boston: Harvard Business Press.

Grenny, J. (2007). "Five crucial conversations for successful projects." *Quality Digest* August, 26–31.

Griffin, T. J. and J. Gustafson. (2007). "Balancing the leadership paradox." *T&D* May, 50–65.

Guaspari, J. (2000). "Solving the mystery of change: The problem isn't next big things; it's next big thing-ism." *Quality Digest* September, 15–24.

Gupta, P. (2007). *Business innovation in the 21st century.* New York: BookSurge Publishing.

Hahn, A. L. (2008). "How poor risk-management techniques contributed to the subprime mess." *CFO* March, 51–58.

Harbour-Falax, L. (2007). "Suppliers: Who will survive?" *Automotive Design and Production* February, 20.

Harrington, H. J. and P. Gupta. (2006). "Six Sigma vs. TQM." *Quality Digest* November, 42–46.

Harrison, M., T. McKinnon, and P. Terry. (2006). *T&D* October, 22–23.

Hartley, D. (2006). "Catalyzing the learning process." *T&D* November, 19–22.

Haynes, C. C. and M. W. Berkowitz. (2007). "What can schools do?" *USA Today* 20 February.

Hayward, M. (2007). "Check your ego for workplace success." *T&D* March, 12.

Hellebusch, S. J. (2009). "Sample quality: Selecting one from many." *Quirk's Marketing Research Review* January, 16–18.

Holman, P., T. Devane, and S. Cady. (2007). *The change handbook,* 2nd ed., rev. and expanded. San Francisco: Berret-Koehler Publishers.

Houston, A., V. Hulton, S. B. Landau, M. Monda, and J. Shettel-Neuber. (1987). Measurement of work processes using statistical process control: Instructor's manual. San Diego, CA: Navy Personnel Research and Development Center. NPRDC Tech. Note 87-17.

Houston, A., J. Shettel-Neuber, and J. P. Sheposh. (1986). Management methods for quality improvement based on statistical process control: A literature and field survey. San Diego, CA: Navy Personnel Research and Development Center. NPRDC Tech. Rep. 86-21.

Hua, Y. (2005). "China's growing appetite." *CFO* November, 75–78.

Huffman, A. (2006). "Lean today, here tomorrow." *Quality Digest* December, 35–39.

Hunt, L., D. Robitaille, and C. Williams. (2008). "The insiders' guide to ISO 9001:2008." *Quality Digest* November, 43–47.

Imler, K. (2006). "Core roles in a strategic quality system." *Quality Progress* June, 57–62.

Iwata, E. (2007). "Businesses grow more socially conscious." *USA Today* 14 February.

Jiang, J., M. Shiu, and M. Tu. (2007). "QFD's evolution in Japan and the West." *Quality Progress* July, 30–38.

Ji-eun, S. (2007). "Honda Accord sedans in record recall." *JoongAng Daily* 21 March.

Johnson, K. (2005). "Six Sigma delivers on-time service." *Quality Progress* December, 57–59.

Johnson, P. (2006). "By the numbers." *Quirk's Marketing Research Review* November, 24–26.

Jones, D. M. (2007). "Merging quality cultures in contract manufacturing." *Quality Progress* September, 41–45.

Kaufman, M. (2006). "Working towards sustainability." *Business Guide* May–June, 54–55.

Kaye, B. and J. Cohen. (2008). "Safeguarding the intellectual capital of baby boomers." *T&D* April, 30–33.

Ketter, P. (2008). "Scouting for leaders." *T&D* March, 28–41.

King, S. (2007). "As new Europe and Asia grow, Britain is losing its influence." *USA Today* 21 May.

Kirchgaessner, S. (2006). "Intellectual property line highlights an Atlantic rift." *Financial Times* 3 November.

Kirkpatrick, J. (2007). "The hidden power of Kirkpatrick's four levels." *T&D* August, 34–37.

Kirscht, R. (2007). "Quality and outsourcing." *Quality Digest* July, 40–43.

Knouse, S. B. (2007). "Building task cohesion to bring teams together." *Quality Progress* March, 49–53.

Krzykowski, B. (2008). "Supply chain management: K'nex success." *Quality Progress* May, 36–43.

Kukor, K. (2009). "A simple plan." *Quality Progress* February, 26–31.

Laff, M. (2006). "Execution is missing in action." *T&D* November, 18.

Laff, M. (2008). "Learning: The role of self-directed learning." *T&D* April, 38–41.

Laflin, L. (2006). "A slight change in the route." *Quirk's Marketing Research Review* October, 40–44.

LaGuardia, D. (2008). "Organizational culture." *T&D* March, 56–63.

Lambet, B. and G. Langvardt. (2008). "So what's in it for the sales team." *T&D* August, 38–41.

Lee, G. (2007). "The hidden cost of poor quality." *Quality Digest* December, 43–53.

Leiss, S. (2005). "Building a better workforce." *CFO* Fall, 20–27.

Leonard, D. (2008). "Strong foundation, solid future." *Quality Progress* March, 30–35.

Leslie, D., D. H. Blackaby, and K. Clark. (1998). *An investigation of racial disadvantage.* Manchester, UK: Manchester University Press.

Levinson, W. A. (1999). "How to design attribute sample plans on a computer." *Quality Digest* July, 45–55.

Lindland, J., F. Love, and R. Sitek. (2000). "Leadership, management and accountability." *Automotive Excellence* Winter, 25–27.

Lindland, J. L. (1999). "Managing change." *Automotive Excellence* Spring, 10–22.

Lippitt, M. (2007). "Fix the disconnect between strategy and execution." *T&D* August, 54–57.

Locher, D. (2007). "In the office: Where Lean and Six Sigma converge." *Quality Progress* October, 54–55.

Locke, A. (2006). "Strategic leadership development." *T&D* December, 53–57.

Lundin, S. C. (2008). *Cats: The nine lives of innovation.* New York: McGraw-Hill.

Malhotra, I. (2006). "Moving and controlling the flow of quality." *Quality Progress* December, 67–69.

Manos, A. (2007). "The benefits of kaizen and kaizen events." *Quality Progress* February, 47–48.

Manos, T. (2006). "Value stream mapping—An introduction." *Quality Progress* June, 64–69.

Marquardt, M. J. (2007). "The power of great questions: Open-ended questions allow for creativity." *T&D* February, 92–93.

Marshak, R. J. (2007). *Covert processes at work.* San Francisco: Berret-Koehler Publishers.

Masters, B. and G. Frazier. (2007). "Project quality activities and goal setting in project performance assessment." *Quality Management Journal* 14(3): 25–35.

Matthews, C. R. (2006). "Linking the supply chain to TQM." *Quality Progress* November, 29–35.

McGee, J. R. (2007). "Creating excellence in a multinational team." *Quality Digest* October, 24–28.

McKee, K. and L. Guthridge. (2007). *Leading people through disasters*. San Francisco: Berret-Koehler Publishers.

McKinley, J. (2008). "Supplying the new domestics." *Automotive Design and Production* March, 16.

Melhotra, I. S. (2006). "Moving and controlling the flow of quality." *Quality Progress* December, 67–69.

Melic, R. (2007). *The rise of the project workforce: Managing people and projects in a flat world*. New York: John Wiley.

Moen, R. D. and T. W. Nolan. (1987). "Process improvement: A step-by-step approach to analyzing and improving a process." *Quality Progress* 20(9): 62–68.

Molnar-Stadler, K. (2002). "Hungary adopts quality tools for public education." *Quality Progress* August, 104–107.

Monroe, D. J. (2006). "Analyzing value streams." *Quality* January, 50–56.

Moore, D. (2000). "Managers, leaders: What's the difference?" *ASQ Communique* Summer, 6–7.

Mora, J. W., A. W. Kaplan, L. M. Wies, and M. A. Atwood. (2008). "Recipe for change." *T&D* March, 42–47.

Morley, I. (2008). "When you hear 'new paradigm,' head for the hills." *Financial Times* 13 June.

Moylan, W. A. (2006). "Leading the virtual project team with trust." *Technology Century* August–September, 36–38.

Murray, M. (2006). "Make the results come alive." *Quirk's Marketing Research Review* October, 58–66.

Nagi, S. and A. Guar. (2000). "The voice of the customer—The starting point of quality." *Quality Times* July, 7–10.

Nancheria, A. (2008). "Retention tension: Keeping high-potential employees." *T&D* March, 18.

Narsh, R. (2002). "Is quick service good service?" *Quirk's Market Research Review* January, 22–25.

Nathan, D. M. (2000). "Multicultural management." *New Straights Times* 29 July.

Neave, H. R. (1990). *The Deming dimension*. Knoxville, TN: SPC Press.

Nugent, P. (2008). "Important considerations in error compensation." *Quality* December, 34–37.

Okes, D. (2006). "Promoting quality in your organization." *Quality Progress* May, 36–41.

Orton, P., D. Beymer, and D. Russell. (2007). "The long and the short of learning: IBM study uncovers link between line length and retention." *T&D* February, 66–69.

O'Sullivan, K. (2006). "Virtue rewarded." *CFO* October, 47–52.

O'Sullivan, K. (2007). "Who's next: Succession planning should be a critical exercise in finance. Too bad so many companies avoid it." *CFO* November, 51–58.

Padma, P., L. S. Ganesh, and C. Rajendran. (2008). "An exploratory sudy of the impact of the capability maturity model on the organizational performance of indian software firms." *Quality Management Journal* 15(2): 20–34.

Palmer, B. (2006). "Selling quality ideas to management." *Quality Progress* May, 27–35.

Palmerino, M. B. (2006). "Qualitative speaking." *Quirk's Marketing Research Review* November, 16–18.

Paton, S. M. (2008). "Are you ready for ISO 9001:2008?" *Quality Digest* November, 56.

Pearse, R. (2007). "Five-pronged approach to climate change challenge." *The Age* 11 May.

Pettit, R. C. (2000). "Data mining: Race for mission-critical info." *Marketing News* 3 January.

Plsek, P. E. (2000). "Creative thinking for surprising quality." *Quality Progress* May, 67–73.

Pollard, D. (2009). "Innovation generator." *T&D* January, 76–77.

Qin, H. and V. R. Prybutok. (2008). "Determinants of customer-perceived service quality in fast-food restaurants and their relationship to customer satisfaction and behavioral intentions." *Quality Management Journal* 15(2): 35–50.

Ramu, G. (2008). "A BOK dedicated to quality in outsourcing is essential in today's global marketplace." *Quality Progress* August, 37–43.

Rappeport, A. (2008). "Game theory versus practice." *CFO* July–August, 35–38.

Reichheld, F. F. (1996). *The loyalty effect*. Boston: Harvard Business School Press.

Reidenbach, R. E. (2007). "Six Sigma, value andcompetitive strategy." *Quality Progress* July, 45–49.

Rhodes, F. H. T. (2006). "Sustainability: The ultimate liberal art." *The Chronicle of Higher Education* 20 October.

Ribeiro, E. (2006). "Differences do matter." *Quirk's Marketing Research Review* November, 50–57.

Ricci, R. (2003). "Move from product to customer centric." *Quality Progress* November, 22–30.

Richarson, S. (2007). "For quality international qualitative, take note of cultural differences." *Quirk's Marketing Research Review* December, 44–49.

Rieder, W. and G. Ganter. (2006). "Collaboration and partnerships among top sources for new, innovative ideas." *Business Guide* May–June, 36–37.

Ritsch, B. (2006). "Breaking the bottleneck: Value-stream mapping helps Lean service industries eliminate waste." *Quality Digest* March, 41–46.

Rooney, J. J., L. N. Vanden Heuvel, and D. K. Lorenzo. (2002). "Reduce human error." *Quality Progress* September, 27–40.

Rowh, M. (2007). "Managing younger workers." *Office Solutions* January, 29–31.

Ruhomally, M. A. (2006). "Never-ending learning." *Quality Progress* December, 64–65.

Ryan, V. (2008). "Banking the subprime crisis." *CFO* March, 49.

Rydholm, J. (2006). "A cooperative effort on cooperation." *Quirk's Marketing Research Review* November, 137, 138.

Sawyer, C. A. (2007). "Play the blame game." *Automotive Design and Production* April, 72.

Sawyer, C. A. (2007). "The innovation situation." *Automotive Design and Production* December, 54–55.

Sawyer, C. A. (2008). "The culture war." *Automotive Design and Production* March, 64.

Schreiber, N. G. (2007). "Don't cross the line: How coaches can master ethical dilemmas in their work." *T&D* February, 60–64.

Schultz, B. (2009). "Out of sight … out of mind." *Quality Progress* February, 20–25.

Schultz, J. R. (2007). "Eight steps to sustain change." *Quality Progress* November, 25–31.

Schultz, J. R. (2008). "Helping ease the transition: Without the right workforce training any process improvement can falter." *Quality Progress* May, 53–58.

Semon, T. T. (2000). "Quality sampling, research means avoiding sloppiness." *Marketing News* 19 June.

Senge, P., B. Smith, N. Kruschwitz, J. Laur, and S. Schley. (2008). "Anatomy of inspiration." *T&D* August, 52–55.

Senturk, D., C. LaComb, R. Neagu, and M. Doganaksoy. (2006). "Detect financial problems with Six Sigma." *Quality Progress* April 41–47.

Shahani, N. N. and A. Raj. (2000). "Customer satisfaction in the service sector: Are companies serious?" *Quality Times* August, 29–30.

Sharma, J. R. and A. M. Rawani. (2007). "Ranking customer's requirements in QFD by factoring in their interrelationship values." *Quality Management Journal* 14(4): 53–60.

Shekoyan, M. (2007). "Using role play and guided imagery for concept generation." *Quirk's Marketing Research Review* December, 32–37.

Sheposh, J. P. and J. Shettel-Neuber. (1986). "Contribution of a multi-method approach to understanding implementation." In *Human factors in organizational design and management*, Vol. 11, edited by O. Brown and H. W. Hendrick. Amsterdam: North-Holland.

Siebert, J. (2007). "Internal customer service: Has it improved?" *Quality Progress* March, 35–40.

Slessareva, L. (2006). "Data use." *Quirk's Marketing Research Review* November, 20–23.

Smith, J. L. (2008). "A recipe for effective leadership." *Quality* December, 42–44.

Snee, R. D. (2006). "Process variation—Enemy and opportunity." *Quality Progress* December, 73–75.

Snee, R. D. (2007). "Methods for business improvement—What's on the horizon." *ASQ Statistics Division: Special Publication* Spring, 9–19.

Stamatis, D. H. (1998). "Needs assessments help you know what to prescribe." *Automotive Excellence* Winter, 18–21.

Su, Q., J. Shi, and S. Lai. (2009). "The power of balance." *Quality Progress* February, 32–37.

Sullivan, K. (2007). "Stuck in a rut: Despite years of investing in IT, outsourcing and process improvements, finance departments still spend too much on routine tasks." *CFO* November, 26.

Sullivan, K. (2007). "Who's next? Succession planning should be a critical exercise in finance. Too bad so many companies avoid it." *CFO* November, 51–58.

Sutherland, J. W. (2006–2007). "Global manufacturing and the sustainability challenge." *Technology Century* December–January, 23–25.

Takacs, S. (2000). "Improve your research through fake data." *Marketing News* 3 January.

Taylor, C. R. (2004). "Retention leadership." *T&D* March, 41–45.

Tazian, V. (2009). "The talent drain." *DBusiness* January–February, 48–51.

Thatchenkery, T. and C. Metzker. (2007). *Appreciative intelligence*. San Francisco: Berret-Koehler Publishers.

Turner, E. and I. Rimanoczy. (2008). "Developing coaching high-impact teams." *T&D* August, 30–37.

Uhlfelder, H. F. (2000). "It's all about improving performance." *Quality Progress* February, 47–54.

Van Mieghem, T. (1995). *Implementing supplier partnerships*. Edgewood Cliffs, NJ: Prentice Hall.

Van Mieghem, T. (1998). "Lessons learned from Alexander the Great." *Quality Progress* January, 41–46.

Vasilagh, G. S. (2003). "Who do you benchmark?" *Automotive Design and Production* March, 26.

Walker, H. F. (2007). "The innovation process and quality tools." *Quality Progress* July, 18–22.

Ward, S., P. Clipp, and S. R. Poling. (2007). "Quality evolves in a Six Sigma world." *Quality* June, 42–45.

Watkins, D. K. (2007). "Conformity or sustainability." *Quality Progress* July, 39–44.

Weatley, M. J. (2007). *Leadership and the new science*, 3rd ed. San Francisco: Berret-Koehler Publishers.

Web, T. and G. Lean. (2007). "The tide is finally turning: Tidal power gains support among PMs." *The Independent on Sunday* 20 May.

Weiss, R. P. (2002). "Crisis leadership." *T&D* March, 28–33.

Whatts, B. and G. B. Dale. (1999). "Small business evaluation and support services: A model from the United Kingdom." *Quality Progress* February, 80–83.

Whitacre, T. (2006). "Map your career through value streams." *Quality Progress* December, 71–72.

Wilson, T. (2008). "A bad climate trade-off." *The Wall Street Journal* 12–14 December.

Winer, B. J. (1971). *Statistical principles in experimental design*, 2nd ed. New York: McGraw-Hill.

Wood, D. C. (2008). "Blurred vision." *Quality Progress* July, 28–33.

Worcester, B. (2007). "Why an ethical business is more likely to produce a success story." *BusinessNews* 18 May.

Wyllie, P. (2006). "One goal, many drivers." *Quirk's Marketing Research Review* October, 46–50.

Zeckhauser, B. and A. Sandoski. (2009). "Using risk to your advantage. *T&D* February, 76–77.

Index

About the Author

Dean H. Stamatis, PhD, ASQC-Fellow, CQE, CMfgE, MSSBB, ISO 9000 Lead Assessor (graduate), is the president of Contemporary Consultants Co. in Southgate, Michigan.

He is a specialist in Management Consulting, Organizational Development, and Quality Science. He has taught Project Management, Operations Management, Logistics, Mathematical Modeling, and Statistics for both graduate and undergraduate levels at Central Michigan University, the University of Michigan, the University of Phoenix, ANHUI University in Bengbu, China, and the Florida Institute of Technology.

With over 30 years of experience in management, quality training, and consulting, Dr. Stamatis has served numerous private sector industries, including (but not limited to) steel, automotive, general manufacturing, tooling, electronics, plastics, food, pharmaceutical, chemical, printing, healthcare, and medical devices, as well as the Navy and the Department of Defense.

He has consulted for such companies as Ford Motor Co., Federal Mogul, GKN, Siemens, Bosch, SunMicrosystems, Hewlett-Packard, GM-Hydromatic, Motorola, IBM, Dell, Texas Instruments, Sandoz, Dawn Foods, Dow Corning Wright, BP Petroleum, Bronx North Central Hospital, Mill Print, St. Claire Hospital, Tokheim, Jabill, Koyoto, SONY, ICM/Krebsoge, Progressive Insurance, B. F. Goodrich, and ORMET, to name just a few.

Dr. Stamatis has created, presented, and implemented quality programs with a focus on Total Quality Management, Statistical Process Control (both normal and short run), Design of Experiments (both classical and Taguchi), Six Sigma (DMAIC and DFSS), Quality Function Deployment, Failure Mode and Effects Analysis, Value Engineering, Supplier Certification, Audits, Reliability & Maintainability, Cost of

Quality, Quality Planning, ISO 9000, QS-9000, ISO/TS 16949, and TE 9000 series.

He has also created, presented, and implemented programs on Project Management, Strategic Planning, Teams, Self-Directed Teams, Facilitator, Leadership, Benchmarking, and Customer Service.

He is a certified Quality Engineer through the American Society of Quality Control, a certified Manufacturing Engineer through the Society of Manufacturing Engineers, a certified Master Black Belt through IABLS, Inc., and he is a graduate of BSI's ISO 9000 Lead Assessor training program.

Dr. Stamatis has written over 70 articles, presented many speeches, and has participated in both national and international conferences on quality. He is a contributing author on several books and the sole author of forty books. His consulting extends across the United States, Southeast Asia, Japan, China, India, Australia, Africa, and Europe. In addition, he has performed over 100 automotive-related audits, 25 pre-assessment ISO 9000 audits, and he has helped several companies attain certification, including Rockwell International – Switching Division (ISO 9001), Transamerica Leasing (ISO 9002), and Detroit Electro Plate (QS-9000).

Dr. Stamatis received his BS/BA degree in Marketing from Wayne State University, his master's degree from Central Michigan University, and his PhD in Instructional Technology and Business/Statistics from Wayne State University.

He is an active member of the Detroit Engineering Society and the American Society for Training and Development, executive member of the American Marketing Association, member of the American Research Association, and a fellow of the American Society for Quality.

For Product Safety Concerns and Information please contact our EU
representative GPSR@taylorandfrancis.com Taylor & Francis Verlag GmbH,
Kaufingerstraße 24, 80331 München, Germany

Printed and bound by CPI Group (UK) Ltd, Croydon, CR0 4YY
08/05/2025
01864418-0001